Galileo and the Equations of Motion

Dino Boccaletti

Galileo and the Equations of Motion

With 26 Figures

 Springer

Dino Boccaletti
Dipartimento di Matematica
University of Rome "La Sapienza"
Rome
Italy

ISBN 978-3-319-36790-3   ISBN 978-3-319-20134-4 (eBook)
DOI 10.1007/978-3-319-20134-4

Springer Cham Heidelberg New York Dordrecht London
© Springer International Publishing Switzerland 2016
Softcover re-print of the Hardcover 1st edition 2016

Printed on acid-free paper

Springer International Publishing AG Switzerland is part of Springer Science+Business Media
(www.springer.com)

To Ruben and Matilde

Preface

During my life as a researcher (a theoretical physicist) and as a teacher (of celestial mechanics, for the last two decades), I always found that the vulgate included in the textbooks of physics not always substantially coincided with what affirmed by the historians of science.

What was taken for granted in the textbooks of physics actually was still a subject of discussion for the scholars of Galileo's works, especially concerning the equations of motion (those we know, and use, under the form of Newton's laws).

From this remark, it came out the idea to "throw a bridge" between the world of the scholars of history of science and the "users" of the textbooks of physics. This bridge is limited to the equations of motion.

Acknowledgments

I wish to thank my wife Maria Grazia and my daughter Chiara for their invaluable help in drawing up and completing this book.

Contents

Introduction... xi
A Preamble.. xi
Notice to the Reader xiv

Part I   The Problem of Motion Before Galileo

1   The Theories on the Motion of Bodies in the Classical Antiquity.....   3
   1.1   Kinematics Among the Greeks..............................   4
   1.2   Dynamics in the Opinion of Aristotle and His Continuators,
         in Greece and in Rome....................................   10
   1.3   After Aristotle in Greece and in Rome....................   18
   Bibliography   22

2   The Theories of Motion in the Middle Ages and in the Renaissance ...   25
   2.1   Preliminary Remarks   25
   2.2   The First Substantial Criticisms to Aristotelian
         Mechanics—Philoponus and Avempace   29
   2.3   The Medieval Kinematics..................................   30
         2.3.1   Gerard of Brussels and the Liber de Motu..........   31
         2.3.2   The Kinematics at Merton College   32
         2.3.3   The Kinematics of the Parisian School   36
   2.4   The Medieval Dynamics....................................   38
         2.4.1   Bradwardine's Dynamics   38
         2.4.2   Dynamics at the Parisian School
                 and the Impetus—Theory...........................   39
   2.5   The Diffusion in Italy of the Ideas of Mertonians
         and of the Parisian Masters   42
   2.6   The Theory of Motion in the XVI Century   44
         2.6.1   Niccolò Tartaglia (1500–1557)—His Life and His Works...   45
         2.6.2   The Mechanics of Giovan Battista Benedetti............   50

2.7   Galileo and the Engineers of the Renaissance   58
Bibliography   60

Part II   Galileo and the Motion

3   **The Young Galileo and the de Motu**   63
3.1   On the Editions of the Dialogue and the Treatise.   63
3.2   The Vicissitudes of the Manuscripts.   65
3.3   The Dialogue   67
3.4   The Treatise   74
3.4.1   The Natural Motions   76
3.4.2   The Circular Motion.   83
3.4.3   The Motion of the Projectiles.   84
3.5   Conclusions   86
3.6   Additional Considerations   89
Bibliography   91

4   **The Inertia Principle**.   93
4.1   The Lex I of Newton   94
4.2   The Inertia in the De Motu.   95
4.3   Inertia in Le Mechaniche and in Dimostrazioni
Intorno Alle Macchie Solari E Loro Accidenti.   98
4.4   The Inertia Principle in the Dialogue and in the Discourses   102
4.5   The Inertia Principle in the "Sixth Day"   109

5   **The Motion of Heavy Bodies and the Trajectory of Projectiles**   117
5.1   An Anticipation   117
5.2   The Motion of Falling Bodies in the "Third Day"
of the Discourses   122
5.2.1   Appendix—An "Unpleasant Incident".   124
5.3   Velocity and Space in the Uniformly Accelerated Motion.   126
5.3.1   Appendix—From the Correspondence
of Paolo Sarpi and Galileo   137
5.4   The Historical Experiment and the Postulate.   137
5.5   The Motion of Projectiles and the Parabolic Trajectory.   148
5.6   A Gloss on the Pendulum.   154
Bibliography   162

6   **Galileo and the Principle of Relativity**   163
6.1   The Relative Motion Before Galileo   164
6.2   How Galileo Expresses the Principle of Relativity   165

Final Considerations.   169

Name Index   173

Introduction

A Preamble

Currently, in the textbooks of physics (at any level of specialization—from the texts for the secondary schools to university handbooks, to research books), a certain number of "results" are credited to Galileo.

Limiting ourselves to the field of dynamics, which we intend to be interested in, one goes from "principles" (inertia principle, principle of relativity—that is so Galilean) to "laws" and, anyhow, relations (motion with constant acceleration, parabolic trajectory of projectiles, inclined planes, isochronism of the small oscillations, etc.).

At present, we all know that in the textbooks, when one says "Tom or Dick equation," he does not want to assert that Tom or Dick had really obtained that equation in that specific form, but that we are in the presence of an equation which comes out from a work, possibly long and hard, unequivocally based on what Tom or Dick had enunciated (guessed or demonstrated).

That is, what we find in the textbooks is just the last formulation, in the formalism and with the notations adopted at present, of an "equation" or of a "law" which is made "to go back" to Tom or Dick.[1]

---

[1] In this connection, it also exists the "Arnold's principle" which states: «if a notion bears a personal name, then this name is not the name of the discoverer».

Do not forget that Galileo was credited, in modern times, with the foundations of dynamics,[2] even if this opinion was questioned.[3] Even if at present the thesis of Pierre Duhem, who maintained that the Parisian School had anticipated Galileo, is no more widely accepted, however, some effects still survive.

We shall discuss onward of this and of the work of the authors of Merton College (Oxford), but we consider necessary here and now to make clear that the assertions of Duhem, in our opinion, do form part of an attitude which should be avoided when one is doing history of science.

One assumes the risk of inventing nonexistent priorities, by interpreting the ancient manuscripts in a contemporaneous perspective, always having an eye to elements which, just read looking at the present, support priorities of theories and discoveries not recognized by the preceding historians.

Therefore, for each of those "results," we must follow "historically" and critically Galileo's path, by bringing out the difficulties, his dependence on forerunners, but also his neat methodological detachment from the latter, he got also by availing himself of experiments, both effectively realized and "ideal."

Of course, we shall run up against still open questions on which the scholars have not drawn generally accepted conclusions, for instance the case of the experiments alluded above. A further difficulty in considering the conclusions reached by Galileo comes from the fact that the terms he used for representing the kinematic magnitudes do not correspond to the concepts and definitions used at present.

Every book entails (or hopes) the existence of a possible type of readers. This one wants to address the persons who are not satisfied with finding enunciations of "laws," "principles," etc., credited to Galileo, but are curious to know what the intellectual debate that culminated in their enunciation has been.

---

[2]On a European level, the clear assertion by Ernst Mach "the founder of dynamics is Galileo" was very important. Mach's work **The Science of mechanics: a critical and historical account of its development** (which contains the quoted assertion) had nine editions (since 1883–1933) and strongly influenced the debate on the subject. We can refer to the English translation by S. MacCormack—The open court publishing, 1960.

[3]At the moment, we limit ourselves to report two excerpts from the well-known preface of Pierre Duhem to the work **Études sur Léonard de Vinci, troisième série—Les précurseurs parisiens de Galilée**—p. VI and pp XIII, XIV. «Lorsque nous voyons la science d'un Galilée triompher du Péripatétisme buté d'un Cremonini, nous croyons, mal informés de l'histoire de la pensée humaine, que nous assistons à la victoire de la jeune Science moderne sur la Philosophie médiéval, obstinée dans son psittacisme; en vérité, nous contemplons le triomphe, longuement préparé, de la science qui est née à Paris au XIVe siècle sur les doctrines d'Aristote et d'Averroès, remises en honneur par la Renaissance italienne.» […] «Cette substitution de la Physique moderne à la Physique d'Aristote a résultè d'un effort de longue durée et d'extraordinaire puissance. Cet effort, il a pris appui sur la plus ancienne et la plus resplendissante des Universités médiévales, sur l'Université de Paris. Comment un parisien n'en serait-il pas fier? Ses promoteurs les plus éminents ont été le picard Jean Buridan et le normand Nicole Oresme. Comment un français n'en éprouverait-il pas un légitime orgueil? Il a résulté de la lutte opiniâtre que l'Université de Paris, véritable gardienne, en ce temp-là, de l'orthodoxie catholique, mena contre le paganisme péripatéticien et néoplatonicien. Comment un chrétien n'en rendrait-il pas grâce a Dieu?»

Besides, history shows that the debate has not always been merely intellectual.

Finally, we would quote a passage of a great historian of science, Paolo Rossi: «… Today the scientists are educated only by reading books written, for the young students, by other scientists. Unlike all the "humanists" who should be ashamed to have never read directly Dante or Shakespeare or Dostoevskij, no physicist and no biologist is obliged or feel morally bound to read the great texts of physics or biology. A few of them, often when aged, read some of these texts, but this is not at all indispensable and after good readings at the secondary schools, one can win the Nobel Prize without having ever opened a scientific textbook prior to the last twenty years» (from Paolo Rossi: **Il tempo dei Maghi—rinascimento e modernità**—Raffaello Cortina Editore, 2006, p. 21).

The reading of a book of physics written some century ago normally requires a considerable intellectual effort from a today researcher. One must immerse himself in an intellectual universe completely different from the present one, and one must adapt himself to a different meaning of terms used (obviously, even when the work is translated in a modern language) and, at last, to recollect the context in which the author was moving.

Not least ambition of this book is to help the hypothetical present-day researcher, mentioned by Paolo Rossi, to read Galileo.

It only remains to explain what the guidelines of our study will be. The rules which is better to specify, though they can be considered obvious, since they will be followed in all this study, will be:

1. All what published (and not subsequently disapproved by Galileo himself) must be considered the "official" thought of the author. Obviously, one must evaluate the chronological order in which the various enunciations have been made and the implicit or explicit constraints due to context. In any case, thinks published and approved represent the prior source.
2. As everybody knows, there exist unpublished works unfinished (or so reputed) as **De Motu** or not directly assembled, by Galileo, as the two known versions of **Le Mecaniche**. They will be used as documents for disclosing the progressive evolution of Galileo's thought from the Pisan period to the Paduan period and the incubation of the new mechanics.
3. The letters. The corpus of the Galilean correspondence is of notable scientific interest. Sometimes, the letters were real scientific essays which were of use, as a channel alternative to publication, for the diffusion of Galileo's thought among the friends, the disciples, and the adversaries. We shall make use of them to integrate, when necessary, the published results, mainly for investigating their origin in the context of the cultural environment of that time. Some results have been communicated through letters long time before the publication.
4. The works of other authors. Along with the letters, there is, obviously, the document which informs us of how Galileo's thought was received by his contemporaries (shared, controversial, rejected, etc.).

In any case, we have preferred to let the texts to speak by themselves, by making the interpretation and the commentary to follow them and not to precede. It is clear that, in so doing, one runs the risk of presenting a piece of writing which resembles an anthology and this is not advisable for an essay. On the other hand, Galileo did not express himself in a concise way, and in trying to summarize his writing, often one is led to act arbitrarily. This possibility is particularly present in the translations. Even Descartes, in the famous letter to Marin Mersenne of October 11 (1638), blamed him for making continuous digressions and not stopping to explain a matter thoroughly.[4] In reality, as it is obvious, the structure of his dialogues was planned for a complete exposition of his thought. As Galileo himself says:

> I have thought it most appropriate to explain these concepts in the form of dialogues, which, not being restricted to the rigorous observance of mathematical laws, make room also for digressions which are sometimes no less interesting than the principal argument.[5]

Notice to the Reader

We are convinced that sometimes the discussion on the thought of the authors (philosophers, scientists) of the past may result misleading if the discussion does not start since the beginning of the original enunciations.

This is, in our opinion, a particularly necessary prerequisite in the case we are going to deal with.

Obviously, all the works from which we have extracted the quoted excerpts were not written in English, and the translations have been subjected to the availability of published English versions. In many cases, we have supplied personal translations.

For Galileo's writings, we have chosen to rely on the translations by Stillman Drake whenever they were available.[6]

---

[4] «… il fait continuellement des digressions et ne s'arrête point à expliquer tout à fait une matière.» – Renè Descartes – Oeuvres de Descartes, 2—Correspondence—Paris: Vrin, 1996.

[5] E. N. VII, p. 30 (Drake, p. 6).

[6] The **Dialogue**, the **Discourses**, the earlier works **De Motu** and **Le Mechaniche** (together with I.E. Drabkin), excerpts from **The Assayer**, **Letters on Sunspots**, and some letters are included in his book **Galileo at work: His Scientific Biography**.

For the **Dialogue** and the **Discourses**, the bibliographical references of the quoted excerpts are dual: the reference to **Galileo Galilei—Dialogue concerning Two Chief world Systems** (The Modern Library—New York, 2001), and to **Galileo Galilei—Two New Sciences** (Wall & Emerson, Inc. Toronto, 2008), respectively, is added to the reference to the national Italian edition (E. N.). In both cases, there is the indication Drake, p. The excerpts from these two works are reprinted with the permission from the publishers. We particularly thank Prof. Byron Wall for his kind permission for free use.

Part I
The Problem of Motion Before Galileo

Chapter 1
The Theories on the Motion
of Bodies in the Classical Antiquity

As we have already said beforehand, our aim is that of investigating a relatively well bounded theme: how the motion of bodies (particularly that further on called "local motion") was dealt with in the centuries preceding the time of Galileo.

Therefore our way of going about things will be deeply different from that of the scholar who makes "history of mechanics". We are not interested in all which has been done in the field of mechanics, but rather in those elements which allow us to trace the (rough) road which led to the laws of motion in the form we know nowadays and whose direct origin we credit to Galileo.

The most obvious thing to do in such a case is that of beginning to explore what happened in the ancient Greece. Usually, one assumes the history of Greek philosophy began with Thales of Miletus (624–545 B.C.) and his followers Anaximander and Anaximenes (the Milesians). Thales is the first philosopher of the group (the pre-Socratics) which includes also Pythagoras and is considered to end with Democritus of Abdera (457–360 B.C.).

We can also say that Thales is the first (to the best of our knowledge) who tries to give a "laic" description of the natural world, i.e. without assuming the involvement of the gods. In his opinion, the Earth floated on the water and the earthquakes were not generated by Poseidon, but by shakes induced by the oscillations of the water.

In the same way, according to Anaximander, the lightnings were due to the wind and the flashes to tearing of the clouds.

It is clear, from the remaining witnesses, that the natural philosophy practised by the pre-Socratics aimed to deal with the great problems: the elements which constituted the world, the existence of celestial bodies and the structure of the cosmos, etc.

Starting from Empedocles, the elements were (starting from the bottom) earth, water, air and fire and all the bodies existing in the sublunary world where composed of them. Except for Leucippus and Democritus, authors of the first atomic

© Springer International Publishing Switzerland 2016
D. Boccaletti, *Galileo and the Equations of Motion*,
DOI 10.1007/978-3-319-20134-4_1

theory, the theory of the four elements will remain valid for the ensuing philosophers (including Aristotle).

The motion in the sense we now mean, that is the shift of a body from one position to another in the space, was not yet object of study. Only the motion of celestial bodies (which lay beyond the sublunary world and moved following a certain periodicity) aroused interest, even on a practical level, for instance for establishing a calendar.

One can assuredly maintain, without fear of exaggerating, that before Aristotle in the Greek science a real dynamic theory did not exist. Obviously, also the term "science" must not be intended in the meaning it has today.

1.1 Kinematics Among the Greeks

As usual nowadays, we have chosen to deal separately with kinematics and dynamics and, given the subject of this book, not to deal with statics (which instead in the Greek world had received a notable increase, suffice it to think of Archimedes, Hero, etc.).

We shall recall this choice many times in the following, at the cost of fastidious repetition, since statics has been the branch of mechanics more extensively studied, and with excellent results, till Galileo; putting it aside obliges us to a "surgical" operation on the works of the authors we shall meet with from time to time.

Inter alia, such an operation is not always straightforward.

For the Greek philosophers, from a certain point on, the problem of motion became a philosophical problem, namely, after the Eleatic school[1] denied its existence asserting it was a spurious appearance.

It fall to Aristotle to deal with this problem in several parts of his works, particularly in the **Physics** and in **On the Heavens**.[2]

By limiting ourselves to extract the mechanical contents, by separating it from the "philosophical" discourse, a quite arbitrary but necessary operation, we have from **On the Heavens**:

> «All natural bodies and magnitudes we hold to be, as such, capable of locomotion; for nature, we say, is their principle of movement. But all movement that is in place, all locomotion, as we term it, is either straight or circular or a combination of these two, which are the only simple movements. And the reason of this is that these two, the straight and the circular line, are the only simple magnitudes.
>
> Now revolution about the centre is circular motion, while the upward and downward movements are in a straight line, "upward" meaning motion away from the centre, and

---

[1]See, infra, the discussion about Zeno's aporias.

[2]Here and afterwards we will refer to the classical translations of R. P. Hardie and R. K. Gaye for **Physics** (Oxford, 1930) and of J. L. Stocks for **On the Heavens** (**De Caelo**), (Oxford, 1930) both available in the Internet Archive http://classics.mit.edu).

"downward" motion towards it. All simple motion, then, must be motion either away from or towards or about the centre.»[3]

Already Plato, in the **Timaeus**, had discussed the notions of "high" and "low", in relation with a spherical world.[4] For Aristotle too, after having confirmed that the centre is the centre of the Earth (that he has established to be spherical on the basis of "proofs"), there are no other elements on the two motions. But now we want to deal with the philosophical argument which exactly asserted the non-existence of what we are talking about (i.e. an argument against motion).

That argument was due to Zeno of Elea (489–431 B.C.). Zeno was a disciple of Parmenides (515–450 B.C.), the founder of the Eleatic School, which was interested in both the foundations of knowledge and the nature of change and becoming.

The argument we want to deal with is well-known and has gone down in history as Zeno's paradox or of Achilles and the tortoise.

The purpose of this digression, since it is really a digression, is that of emphasizing how much, in the treatment of a problem, the definitions of physical quantities are essential for obtaining results making sense.

Zeno's debate was against the plurality and the change and negation of motion had nothing to do with kinematics, it was only a way to use an example for confirming the negation of the reality of space, time and plurality.

Let us see how Aristotle relates Zeno's arguments:

«… Zeno's arguments about motion, which cause so much disquietude to those who try to solve the problems that they present, are four in number.

The first asserts the nonexistence of motion on the ground that that which is in locomotion must arrive at the halfway stage before it arrives at the goal. This we have discussed above.

The second is the so-called "Achilles", and it amounts to this, that in a race the quickest runner can never overtake the slowest, since the pursuer must first reach the point whence the pursued started, so that the slower must always hold a lead. …»[5]

We are not interested in seeing how Zeno's ideas have been discussed and refuted by the other philosophers over the centuries (from Aristotle to Bertrand Russel).

Let us see instead the "mathematical" confutation we can often find nowadays in textbooks of mathematics, as an exercise of application of the convergence of series.[6]

---

[3]**On the Heavens**, I 2-268b.

[4]See Plato: **Timaeus**, XXVI, 63. The reader can refer to **Timaeus** translated by Benjamin Jowet available in http://classics.mit.edu/Plato/timaeus.html.

[5]Aristotle: **Physics**, VI, 9 239b. As in the case of several other ancient phylosophers, the original works of Zeno are missing so one must refer to the excerpts quoted by other phylosophers in their commentaries. Besides Aristotle, also Plato and Simplicius are sources for knowing Zeno's work.

[6]As a historical curiosity we mention that the first suggestion for solving Zeno's paradox by means of the sum of a convergent geometrical series goes back to Grégoire de Saint-Vincent in the work **Opus geometricum quadraturae circuli et sectionum coni decem libris comprehensum—Antverpiae, 1647.**

Let us schematize the problem as follows.

Consider the motion taking place on a straight line where Achilles starts from point A. If we indicate with t_1 the instant in which Achilles walks the distance d, we shall have $S_A(t_1) = d$, having called S_A the space walked by A. But in the meanwhile tortoise T will have reached another point which we shall call $S_T(t_1)$; when, at instant t_2, A will have reached the position formerly occupied by T, we shall have $S_A(t_2) = S_T(t_1)$.

Going on, at the subsequent instants we shall have a situation we can express with $S_A(t_{n+1}) = S_T(t_n)$, where index n labels the running instant. If, for the sake of simplicity, we assume that both A and T move with constant velocities V_A and V_T (obviously with $V_A > V_T$), we shall have $S_A(t) = V_A \cdot t, S_T(t) = d + V_T \cdot t$ and then, in general $S_A(t_{n+1}) = V_A \cdot t_{n+1} = S_T(t_n) = V_T t_n + d$, wherefrom

$$t_1 = \frac{d}{V_A}, \ldots, t_{n+1} = \frac{V_T \cdot t_n + d}{V_A}.$$

By interpreting in this formalism Zeno's idea, we should have equal to infinity the sum of infinite time intervals $t_{n+1} - t_n$. That is, writing such a sum as the series

$$\sum_{k=1}^{\infty} (t_{k+1} - t_k),$$

the series should diverge. Let us build it. From the relations written above, we have: $t_1 = \frac{d}{V_A}, t_2 = \frac{V_T t_1 + d}{V_A}, t_3 = \frac{V_T t_2 + d}{V_A}, \ldots$ from which $t_2 - t_1 = \frac{V_T}{V_A} t_1, t_3 - t_2 = (\frac{V_T}{V_A})^2 \cdot t_1$ and, going on, $t_{n+1} - t_n = (\frac{V_T}{V_A})^n \cdot t_1$.

Therefore, the series will be

$$\sum_{k=1}^{\infty} (t_{k+1} - t_k) = t_1 \cdot \sum_{K=1}^{\infty} \left(\frac{V_T}{V_A}\right)^K.$$

Being $V_T < V_A$, it will be $\left|\frac{V_T}{V_A}\right| < 1$ and then we are in presence of the convergent geometric series having the sum

$$\sum_{K=1}^{\infty} \left(\frac{V_T}{V_A}\right)^K = \frac{1}{1 - \frac{V_T}{V_A}} = \frac{V_A}{V_A - V_T}$$

wherefrom the sum of the time intervals will be $\frac{V_A}{V_A - V_T} \cdot t_1 = \frac{d}{V_A - V_T}$.

Therefore, Achilles reaches the tortoise in a time $\frac{d}{V_A - V_T}$.

Let us try now to behave like everyman who does not doubt that Achilles will reach immediately the tortoise in a certain time t which can be determined simply by putting $S_A(t) = S_T(t) \Rightarrow V_A \cdot t = d + V_T \cdot t$, wherefrom $t = \frac{d}{V_A - V_T}$.

It cannot be passed unnoticed to the reader the fact that, both the hypothetical mathematics student and the man in the street, have considered that the velocity of a moving body can be obtained by dividing the space covered for the time passed (at least in the case of uniform motion).

This, which is for us an elementary notion of kinematics (that is the definition of velocity) required more than twenty centuries after the time of Zeno to come into use!

We can add that the demonstration with the support of a convergent series dates from the time when mathematicians succeeded in demonstrating that the sum of infinite (finite) terms may also result finite. This mathematical solution of Zeno's paradox elegantly closed the question, but we must instead take care of grasping how one could arrive at the solution of the man in the street.

Then we must follow, starting from its beginning, the slow evolution of that branch of mechanics named kinematics. What we would succeed in doing in the following is to separate the kinematics from the dynamics in order to be able of better evaluating the progress occurred in each of the two branches, even if the historical development has not taken place in the order in which they are arranged in the modern texts of rational mechanics.

«On a donné le nom de cinématique à l'étude du mouvement, indépendamment des causes qui le produisent, c'est-à-dire indépendamment des forces. Il apparaît comme logique d'étudier d'abord les forces indépendamment du mouvement, ce qui est l'objet de la statique, puis le mouvement indépendamment des forces, c'est-à-dire la cinématique, avant d'aborder la dinamique, c'est-à-dire l'étude des relations entre les forces et le mouvement.

En fait, les historiens de la Mécanique ne font guère mention des origines et des progrès de la cinématique et il semble bien que, jusqu'à une époque relativement récente, l'histoire de la cinématique se soit confondue, soit avec celle de la géométrie, soit avec celle de la dynamique.

Il y a cependent interêt à rechercher comment les principes fondamentaux de la cinématique se sont dégagés peu a peu de l'empirisme primitif pour arriver à constituer une doctrine cohérente».[7]

Indeed only recently and, we would say, more recently of what Borel says for the kinematics in a whole, one has begun to study how the concept of velocity has been intended over the centuries, starting from the Greek philosophers till the birth of modern physics.[8]

It seems to us that the condition which, along the centuries, has prevented the natural philosophers from having a correct (in the modern sense) definition of velocity depends on the lack of the suitable mathematical algorithm to be applied to the description of physical phenomena.

It was obvious that, for characterizing the swiftness of a motion, it was necessary to make comparisons between distances and times spent to cover them, but the only available mathematical algorithm was that of proportions, earlier

---

[7]Émile Borel: **L'évolution de la mécanique**—Flammarion, 1943—p. 45.

[8]A very interesting paper is due to Pierre Souffrin: **Sur l'histoire du concept de vitesse d'Aristote à Galilée**—Medioevo—Rivista di studi della filosofia medievale, vol. XXIX (2004), pp. 99–133. In this paper, the author introduces the concept of holistic velocity to represent the meaning of the term "velocity" used by ancient philosophers. He denotes it with /velocitas/.

studied by Eudoxus (408–335 B.C.) then codified in the fifth book of Euclides' (323–286 B.C.) **Elements**, based on the equality of ratios between homogeneous magnitudes.

Souffrin says

«Les définitions, ou des énoncés équivalents à des définitions de/velocitas/sont remarquablement rares dans les textes anciens.

Il en est de même des définitions de l'uniformité ou de la nonuniformité du movement, dont le rapport aux acceptions de/velocitas/est historiquement significatif. Il faut voir là, sans doute, un reflet du caractère fondamental, disons primordial, des concepts en question.

On peut penser que des usages vernaculaires des notions d'uniformité et de vitesse ont largement précédé l'apparition d'une quelconque nécessité de les formalizer et l'élaboration d'outils conceptuels en rendant possibles des expreessions mathématiques.

Dans ces conditions, les termes correspondants semblent appartenir le plus souvent, dans les textes anciens, à une sorte de métalangage qui peut servir à définir mais qu'il ne convient pas de définir.»

Then, going on to discuss the passages of Aristotle's **Physics** where the uniform motion is dealt with (the uniform motion is the only type of motion of which a hint of a description is given using the proportions), he points out their indeterminateness in the one hand and their obviousness on the other hand.

Let us see Aristotle directly:

«The difference that makes a motion irregular is sometimes to be found in its path: thus a motion cannot be regular if its path is an irregular magnitude, e.g. a broken line, a spiral, or any other magnitude that is not such that any part of it taken at random fits on to any other that may be chosen.

Sometimes it is found neither in the place nor in the time nor in the goal but in the manner of the motion: for in some cases the motion is differentiated by quickness and slowness: thus if its velocity is uniform a motion is regular, if not it is irregular.»[9]

And more

«In all cases where a thing is in motion with uniform velocity it is clear that the finite magnitude is traversed in a finite time. For if we take a part of the motion which shall be a measure of the whole, the whole motion is completed in as many equal periods of the time as there are parts of the motion.

Consequently, since these parts are finite, both in size individually and in number collectively, the whole time must also be finite: for it will be a multiple of the portion, equal to the time occupied in completing the aforesaid part multiplied by the number of the parts.»[10]

Hanging on to our assumption of discussing separately the contributions of kinematics and dynamics, ranging from Aristotle to the Hellenistic age, we meet only two authors who somehow define the uniform motion (obviously, only this).

---

[9]**Physics**, V, 228b 27.

[10]Ibidem, VI 237b 25.

The first one is Autolicus of Pitane, who lived about 300 B.C. (just prior to Euclid), who in his work **On the moving Sphere**[11] deals with the motion of uniform rotation. Let us see the terms:

I. When a sphere rotates with uniform velocity around its axis, all points on its surface, not lying on the axis, describe parallel circles having their poles coinciding with the poles of the sphere and their planes perpendicular to the axis.

II. All those points describe similar arcs on their parallels in equal times.

III. Vice versa similar arcs are described in equal times.»[12]

Loria remarks:

«The propositions mentioned above are so simplistic and so scantily interesting for the geometer that certainly one would not have thought to collect them in a particular treatise if one had not seen that, through them, it was possible to explain some celestial phenomena.»[13]

The second author, far more important, is the great Archimedes (287–212 B.C.) who defines the uniform motion in two propositions contained in the work **On Spirals**.[14]

In this work Archimedes introduces and studies the curve which later on was called "Archimedes spiral".

It can be obtained in this way:

«… If a straight line of which one extremity remains fixed be made to revolve at a uniform rate in a plane until it returns to the position from which it started, and if, at the same time as the straight line revolves, a point move at a uniform rate along the straight line, starting from the fixed extremity, the point will describe a spiral in the plane.»[15]

Therefore, for Archimedes, it is necessary above all to show what one means, precisely, when speaking of a motion at constant velocity (the Greek term, to the letter, means "equally fast to itself").

Let us see the two propositions:

«Proposition 1. If a point moves at a uniform rate along any line, and two lengths be taken on it, they will be proportional to the times of describing them.»[16]

That is, if s_1 and s_2 are the lengths of the two segments and t_1 and t_2 the times spent to cover them, one has $s_1 : s_2 = t_1 : t_2$, which is the classical Euclides proportion.

---

[11]To the best of our knowledge, an English translation of this work does not exist. We translate from the Italian translation of Gino Loria's **Le Scienze esatte nell'antica Grecia**—Hoepli, 1914.

[12]Ibidem, p. 499.

[13]Ibidem, p. 500.

[14]For Archimede's work, see T. L. Heath: **The Works of Archimedes**—Cambridge University Press—1897, reprinted by Dover, 2002. Also E. J. Dijksterhuis: **Archimedes**—Princeton University Press, 2014 (the original edition was published in 1938).

[15]See Heath, op.cit., p. 154.

[16]Ibidem, p. 155.

Finally:

«Proposition II. If each of two points on different lines respectively move along them each at a uniform rate, and if lengths be taken, one on each line, forming pairs, such that each pair are described in equal times, the lengths will be proportional.»[17]

Whereas in Proposition I only one uniform motion was considered, demonstrating the proportionality between the spaces covered and the times taken to cover them, in Proposition II two uniform motions are considered and the spaces covered in each of the two motions in equal times are compared.

In this way, Archimedes, by means of the algorithm of proportions, has established the definition of uniform (rectilinear) motion.

The Euclidean algorithm will maintain the predominance until the time of Newton. It will be the turn of Euler (1707–1783) to free the mechanics «from the chains of the synthetic demonstration, transforming it in an analytic science.»[18]

In this regard, E. T. Bell deals out a second dose by saying that Newton's **Principia** could even have been written by Archimedes whereas Euler's **Mechanics** could not have been written by a Greek.

In fact, in the work **Mechanica sive Motus Scientia analytice exposita**,[19] published in 1736 (ten years after the third edition of **Principia** and nine after Newton's death), Euler makes use of the newborn infinitesimal analysis and rids himself from the geometrical methods of Greek origin.

Also in the simple definition of the uniform motion one does not use the proportion algorithm anymore and, finally, the velocity is obtained by dividing the space covered by the time taken to cover it.

In fact «Corollary 5.30. In a similar way the velocity will be expressed by the space covered divided by the time and also the space itself by the product of time for the velocity.»[20]

1.2 Dynamics in the Opinion of Aristotle and His Continuators, in Greece and in Rome

As Heath says,[21] Aristotle, even though not being a mathematician by profession, was well-informed about the mathematics of his time, at least when one was dealing with elementary mathematics. Therefore we have not to expect, in the case of

---

[17]Ibidem, p. 155.

[18]E. T. Bell: **Men of Mathematics**—Simon & Schuster, 1937, Chap. IX.

[19]**Mechanica sive Motus Scientia analytice exposita Auctore Leonhardo Eulero**—Petropoli, A. 1736.

[20]**Mechanica ...** op. cit. p. 11. «Corollarium 5.30. Similiter celeritas poterit exprimi per spatium percursum divisum per tempus, atque spatium etiam ipsum per factum ex tempore in celeritatem.».

[21]See: T. L. Heath: **Mathematics in Aristotle**—Oxford, 1949, p. 1 (He was evidently not a mathematician by profession, but he was abreast of the knowledge of his days as far as elementary mathematics is concerned; ...).

dynamics, to find terms which foreshadow a mathematical physics to come. One has to do, practically in most cases, with assertions which are based on an observation of natural processes carried out following the common sense.

A comment by Koyré about this topic:

«La Physique aristotélicienne est essentiellement non mathèmatique, et on ne peut la mathématiser (en la présentant, par exemple, comme fondée sur le principe: vitesse proportionelle à la force et inversement proportionelle à la éxistence, proportionnalité qui n'est qu'une *suite* des principes aristotéliciens), sans fausser l'esprit.»[22]

If we should classify Aristotle (following the modern attributions) from the point of view of his activity in the scientific research, we would say that he was a naturalist, certainly not a physicist.

But his predominance (jointly with that of his disciples and continuators) in the philosophical field made it sure that the things he maintained both in the field of mechanics and on the structure of the universe were considered a sound and indisputable basis on which every future research should be grafted. As we have already mentioned, Aristotle dealt with the motion in several parts of his works and, since the purpose of his philosophy was that of investigating the "causes" of the things, when applied to the motion this led to build a theory of dynamics, «c'est-à-dire l'étude des relations entre les forces et le mouvement», as concisely asserts the passage of Borel which footnote 7 is referred to.

As a rule, for interpreting the nature, he maintains:

«... Obviously then it would be better to assume a finite number of principles. They should, in few, be as few as possible, consistently with proving what has to be proved. This is the common demand of mathematicians, who always assume as principles things finite either in kind or in number.»[23]

In Aristotle's opinion, there exist two kinds of motion: the natural motion and the violent motion (under constriction):

«Now all things rest and move naturally and by constraint. A thing moves naturally to a place in which it rests without constraint, and rests naturally in a place to which it moves without constraint. On the other hand, a thing moves by constraint to a place in which it rests by constraint, and rests by constraint in a place to which it moves by constraint. Further, if a given movement is due to constraint, its contrary is natural. If then, it is by constraint that earth moves from a certain place to the centre here, its movement from here to there will be natural, and if earth from there rests here without constraint, its movement hither will be natural. And the natural movement in each case is one.»[24]

The direction of the rectilinear natural motion may be upward or downward and the sense is exclusively determined by the nature of the body without any relation with its volume. The circular motion in accordance with nature, on the contrary, belongs to a particular body: the ether.

---

[22]Alexandre Koyré: **À l'aube de la science classique**—Paris, Hermann, 1939, p. 11.

[23]**On the Heavens**, III, 4, 302.

[24]Ibidem, I, 8, 276 a.

Let Aristotle speak directly:

«Bodies are either simple or compounded of such; and by simple bodies I mean those which possess a principle of movement in their own nature, such as fire and earth with their kinds, and whatever is akin to them. Necessarily, then, movements also will be either simple or in some sort compound simple in the case of the simple bodies, compound in that of the composite and in the latter case the motion will be that of the simple body which prevails in the composition. Supposing, then, that there is such a thing as simple movement, and that circular movement is an instance of it, and that both movement of a simple body is simple and simple movement is of a simple body (for if it is movement of a compound it will be in virtue of a prevailing simple element), then there must necessarily be some simple body which revolves naturally and in virtue of its own nature with a circular movement.

By constraint, of course, it may be brought to move with the motion of something else different from itself, but it cannot so move naturally, since there is one sort of movement natural to each of the simple bodies. Again, if the unnatural movement is the contrary of the natural and a thing can have no more than one contrary, it will follow that circular movement, being a simple motion, must be unnatural, if it is not natural, to the body moved. If then the body, whose movement is circular, is fire or some other element, its natural motion must be the contrary of the circular motion.

But a single thing has a single contrary; and upward and downward motion are the contraries of one another. If, on the other hand, the body moving with this circular motion which is unnatural to it is something different from the elements, there will be some other motion which is natural to it. But this cannot be. For if the natural motion is upward, it will be fire or air, and if downward, water or earth. Further, this circular motion is necessarily primary. For the perfect is naturally prior to the imperfect, and the circle is a perfect thing. This cannot be said of any straight line: -not of an infinite line; for, if it were perfect, it would have a limit and an end: nor of any finite line; for in every case there is something beyond it, since any finite line can be extended. And so, since the prior movement belongs to the body which naturally prior, and circular movement is prior to straight, and movement in a straight line belongs to simple bodies fire moving straight upward and earthy bodies straight downward towards the centre since this is so, it follows that circular movement also must be the movement of some simple body.

For the movement of composite bodies is, as we said, determined by that simple body which preponderates in the composition. These premises clearly give the conclusion that there is in nature some bodily substance other than the formations we know, prior to them all and more divine than they.

But it may also be proved as follows. We may take it that all movement is either natural or unnatural, and that the movement which is unnatural to one body is natural to another as, for instance, is the case with the upward and downward movements, which are natural and unnatural to fire and earth respectively. It necessarily follows that circular movement, being unnatural to these bodies, is the natural movement of some other. Further, if, on the one hand, circular movement is natural to something, it must surely be some simple and primary body which is ordained to move with a natural circular motion, as fire is ordained to fly up and earth down. If, on the other hand, the movement of the rotating bodies about the centre is unnatural, it would be remarkable and indeed quite inconceivable that this movement alone should be continuous and eternal, being nevertheless contrary to nature.

At any rate the evidence of all other cases goes to show that it is the unnatural which quickest passes away. And so, if, as some say, the body so moved is fire, this movement is just as unnatural to it as downward movement; for any one can see that fire moves in a straight line away from the centre.

On all these grounds, therefore, we may infer with confidence that there is something beyond the bodies that are about us on this earth, different and separate from them; and that the superior glory of its nature is proportionate to its distance from this world of ours. In consequence of what has been said, in part by way of assumption and in part by way of proof, it is clear that not everybody either possesses lightness or heaviness.

As a preliminary we must explain in what sense we are using the words "heavy" and "light", sufficiently, at least, for our present purpose: we can examine the terms more closely later, when we come to consider their essential nature.

Let us then apply the term "heavy" to that which naturally moves towards the centre, and "light" to that which moves naturally away from the centre. The heaviest thing will be that which sinks to the bottom of all things that move downward, and the lightest that which rises to the surface of everything that moves upward. Now, necessarily, everything which moves either up or down possesses lightness or heaviness or both-but not both relatively to the same thing: for things are heavy and light relatively to one another; air, for instance, is light relatively to water, and water tight relatively to earth.

The body, then, which moves in a circle cannot possibly possess either heaviness or lightness. For neither naturally nor unnaturally can it move either towards or away from the centre. Movement in a straight line certainly does not belong to it naturally, since one sort of movement is, as we saw, appropriate to each simple body, and so we should be compelled to identify it with one of the bodies which move in this way. Suppose, then, that the movement is unnatural. In that case, if it is the downward movement which is unnatural, the upward movement will be natural; and if it is the upward which is unnatural, the downward will be natural For we decided that of contrary movements, if the one is unnatural to anything, the other will be natural to it. But since the natural movement of the whole and of its part of earth, for instance, as a whole and of a small clod-have one and the same direction, it results, in the first place, that this body can possess no lightness or heaviness at all (for that would mean that it could move by its own nature either from or towards the centre, which, as we know, is impossible); and, secondary, that it cannot possibly move in the way of locomotion by being forced violently aside in an upward or downward direction. For neither naturally nor unnaturally can it move with any other motion but its own, either itself or any part of it, since the reasoning which applies to the whole applies also to the part. It is equally reasonable to assume that this body will be un-generated and indestructible and exempt from increase and alteration, since everything that comes to be comes into being from its contrary and in some substrate, and passes away likewise in a substrate by the action of the contrary into the contrary, as we explained in our opening discussions.

Now the motions of contraries are contrary. If then this body can have no contrary, because there can be no contrary motion to the circular, nature seems justly to have exempted from contraries the body which was to be un-generated and indestructible. For it is in contraries that generation and decay subsist.»[25]

Therefore, the ether, or quintessence, is this particular constituent (beyond the sublunary world where the four elements are) of the celestial spheres where the eternal circular motion is operative. We are interested only in "local motion".

---

[25]Ibidem, I, 2–3, 269 a–270 a.

In the case of natural motion, we shall have only rectilinear motions, upward or downward.

«By absolutely light, then, we mean that which moves upward or to the extremity, and by absolutely heavy that which moves downward or to the centre. By lighter or relatively light we mean that one, of two bodies endowed with weight and equal in bulk, which is exceeded by the other in the speed of its natural downward movement.»[26]

Moreover:

«A given weight moves a given distance in a given time; a weight which is as great and more moves the same distance in a less time, the times being in inverse proportion to the weights. For instance, if one weight is twice another, it will take half as long over a given movement.»[27]

«Secondly, like the upward movement of fire, the downward movement of earth and all heavy things makes equal angles on every side with the earth's surface: it must therefore be directed towards the centre. Whether it is really the centre of the earth and not rather that of the whole to which it moves, may be left to another inquiry, since these are coincident.»[28]

In a passage, Aristotle seems to guess a thing which remind us of what will be the Archimedes principle. But it is a blunder, not least because contradicted by the author himself:

«It is due to the properties of the elementary bodies that a body which is regarded as light in one place is regarded as heavy in another, and vice versa. In air, for instance, a talent's weight of wood is heavier than a mina of lead, but in water the wood is the lighter. The reason is that all the elements except fire have weight and all but earth lightness.

Earth, then, and bodies in which earth preponderates, must needs have weight everywhere, while water is heavy anywhere but in earth, and air is heavy when not in water or earth. In its own place each of these bodies has weight except fire, even air. Of this we have evidence in the fact that a bladder when inflated weighs more than when empty.

A body, then, in which air preponderates over earth and water, may well be lighter than something in water and yet heavier than it in air, since such a body does not rise in air but rises to the surface in water.»[29]

And at last, on the motion under constriction:

«If everything that is in motion with the exception of things that move themselves is moved by something else, how is it that some things, e.g. things thrown, continue to be in motion when their movent is no longer in contact with them? If we say that the movent in such cases moves something else at the same time, that the thrower e.g. also moves

---

[26]Ibidem, IV, 1, 308 a.

[27]Ibidem, I, 6, 274 a.

[28]Ibidem, IV, 4, 312 a. For a unanimous recognition of the commentators, here is meant that the expression "equal angles" must be intended in the sense that they form angles of 90° with the local tangent to the terrestrial surface. Therefore, in different places, the heavy bodies do not fall parallel but their trajectories converge in the centre of the Earth which, as we know, Aristotle has established to be spherical.

[29]Ibidem, IV, 4, 311 b.

the air, and that this in being moved is also a movent, then it would be no more possible for this second thing than for the original thing to be in motion when the original movent is not in contact with it or moving it: all the things moved would have to be in motion simultaneously and also to have ceased simultaneously to be in motion when the original movent ceases to move them, even if, like the magnet, it makes that which it has moved capable of being a movent.

Therefore, while we must accept this explanation to the extent of saying that the original movent gives the power of being a movent either to air or to water or to something else of the kind, naturally adapted for imparting and undergoing motion, we must say further that this thing does not cease simultaneously to impart motion and to undergo motion: it ceases to be in motion at the moment when its movent ceases to move it, but it still remains a movent, and so it causes something else consecutive with it to be in motion, and of this again the same may be said.

The motion begins to cease when the motive force produced in one member of the consecutive series is.at each stage less than that possessed by the preceding member, and it finally ceases when one member no longer causes the next member to be a movent but only causes it to be in motion. The motion of these last two—of the one as movent and of the other as moved—must cease simultaneously, and with this the whole motion ceases. Now the things in which this motion is produced are things that admit of being sometimes in motion and sometimes at rest, and the motion is not continuous but only appears so: for it is motion of things that are either successive or in contact, there being not one movent but a number of movents consecutive with one another: and so motion of this kind takes place in air and water.

Some say that it is "mutual replacement": but we must recognize that the difficulty raised cannot be solved otherwise than in the way we have described. So far as they are affected by "mutual replacement", all the members of the series are moved and impart motion simultaneously, so that their motions also cease simultaneously:».[30]

The modern dynamics (that which we call classical) deals with the motion of bodies in space and time and establishes its fundamental laws by considering the motion in empty space.

It was not so in Aristotle's opinion. He denied the existence of void and maintained that the motion happened in the "place", which is the "category" to which he dedicates a wide room in the fourth book of **Physics**.

«Place is thought to be something important and hard to grasp, both because the matter and the shape present themselves along with it, and because the displacement of the body that is moved takes place in a stationary container, for it seems possible that there should be an interval which is other than the bodies which are moved. The air, too, which is thought to be incorporeal, contributes something to the belief: it is not only the boundaries of the vessel which seem to be place, but also what is between them, regarded as empty. Just, in fact, as the vessel is transportable place, so place is a non-portable vessel.

So when what is within a thing which is moved, is moved and changes its place, as a boat on a river, what contains plays the part of a vessel rather than that of place. Place on the other hand is rather what is motionless: so it is rather the whole river that is place, because as a whole it is motionless.

---

[30]**Physics**, VIII, 10, 266 b. We note that the concept of antiperistasis, introduced by Empedocles, had been already used by Plato in the **Timaeus** (XXXVII, 79–80), to explain the phenomenon of respiration.

Hence we conclude that the innermost motionless boundary of what contains is place.»[31]

The last phrase will be the definition universally used in the Middle Ages in the Latin version "terminus continentis immobilis primus".

From this definition of "place" obviously also derives the impossibility of the existence of void. In fact, if one characterizes the void as a place where there is nothing, this is a contradiction in terms with the definition of place as a "terminus continentis".

In any case, Aristotle reaffirms the non-existence of void in several parts of his work with a variety of arguments.

As we have mentioned at the beginning, Aristotle has a limited knowledge of mathematics, so to say, at a mean level (for instance, he never alludes to the classical problems of the Greek mathematics—trisection of an angle, duplication of the cube, rectification of the circumference).

Then it is no use to go and look for mathematical elements into his definitions and demonstrations regarding mechanical problems. As Giorgio De Santillana says:

«He is the type of scientist who remains always closely bound to sensible, experienced reality. For him velocity is proportional to effort, as anyone knows who has to quicken his pace or to push a wheelbarrow; space is the collection of "places" of things, in a way which would remind us of the shipping room in a department store.»[32]

And, at last, G. E. R. Loyd:

«He has often been condemned for assuming that the speed of a freely falling body varies directly with its weight. But the fact is that in air heavier bodies do fall more rapidly than lighter ones of the same shape and size, although this is not true in a vacuum. He was correct in assuming that there is some relationship between weight and speed in motion that takes place through a medium, although the relationship is not a simple one of direct proportion. Similarly it is obviously true that motion through a dense medium is generally slower than through a rare one, but again he oversimplified the relationship in treating it as one of a direct proportion.

The main shortcoming of Aristotle's dynamics is not so much a failure to pay attention to the data of experience, as a failure to carry abstraction far enough.»[33]

In the first book of **On the Heavens**, Aristotle mentioned the acceleration of the heavy bodies in free fall:

«This conclusion that local movement is not continued to infinity is corroborated by the fact that earth moves more quickly the nearer it is to the centre, and fire the nearer it is to the upper place. But if movement were infinite speed would be infinite also; and if speed then weight and lightness. For as superior speed in downward movement implies superior weight, so infinite increase of weight necessitates infinite increase of speed. Further, it is not the action of another body that makes one of these bodies move up and the other

---

[31]Ibidem, IV, 4, 212 a.

[32]G. de Santillana: **The Origins of Scientific Thought**—The University of Chicago Press, 1961—Chap. 13.

[33]Geoffrey E. R. Lloyd: **Early Greek Science**—Chatto & Windus Ltd, London—1970, Chap. 8.

down; nor is it constraint, like the "extrusion" of some writers. For in that case the larger the mass of fire or earth the slower would be the upward or downward movement; but the fact is the reverse: the greater the mass of fire or earth the quicker always is its movement towards its own place.»[34]

The subject will be taken up by his continuator Strato of Lampsacus (who will direct the Liceum of Aristotle from about 286 to 268, succeeding to Theophrastus).

Strato, known as "the physicist", dealt with gravity and void and took up original positions, with respect to Aristotle, on these themes. Unfortunately his works have gone missing and of his treatise **On Motion** one has only fragments quoted by Simplicius (about 490-about 560 A.D.) in his **Commentary on Aristotle's Physics**.

In fact, concerning the acceleration, Simplicius remarks:

«It is universally asserted as self-evident that bodies moving naturally to their natural places undergo acceleration. As to the cause of such acceleration some say, and reasonably, that the bodies are endowed with greater force as they approach more closely to their proper wholeness, that is, as they achieve greater perfection of form.

Others, again, say that it is the quantity of intervening air that impedes bodies moving upward or downward, until these bodies approach their natural places, when only a small amount of the medium is left to be traversed.

But few adduce any proof of the fact itself, that when bodies moving naturally are near their natural places they move more swiftly.

It may therefore not be out of place to set forth the indications [of acceleration] given by Strato, the Physicist. For in his treatise On Motion, after asserting that a body so moving completes the last stage of its trajectory in the shortest time, he adds: "In the case of bodies moving through the air under the influence of their weight this is clearly what happens. For if one observes water pouring down from a roof and falling from a considerable height, the flow at the top is seen to be continuous, but the water at the bottom falls to the ground in discontinuous parts. This would never happen unless the water traversed each successive space more swiftly." By "this" Strato means the breaking up of the continuity of the object as it approaches the ground. Strato also adduces another argument, as follows:

"If one drops a stone or any other weight from a height of about an inch, the impact made on the ground will not be perceptible, but if one drops the object from a height of a hundred feet or more, the impact on the ground will be a powerful one. Now there is no other cause for this powerful impact. For the weight of the object is not greater, the object itself has not become greater, it does not strike a greater space of ground, nor is it impelled by a greater [external force]. It is merely a case of acceleration. And it is because of this acceleration that this phenomenon and many others take place."»[35]

At last, we quote Loyd's opinion who remarks that «the importance of this text does not consists in what Strato strives to demonstrate (the existence of the acceleration) but instead in the method he follows.»

---

[34]**On the Heavens** I, 8, 277 a–b.

[35]See **A Source Book in Greek Science**, Morris R. Cohen and I.E. Drabkin Eds.—Harvard University Press, 1966—pp. 211–212.

In fact, in Strato's discourse there is an appeal to experience; even if one has not to do with an experiment really prepared for that purpose, however there he describes phenomena which everyone can see or has seen and then can serve as "experimental proof".

Nevertheless, one turns to experiments (this time effectively arranged to the purpose) for demonstrating the existence of the void (in disagreement with Aristotle).

The exposition of the reasonings, with the description of the experimental apparatus, can be found in the **Pneumatics** of Hero (the period of his life is uncertain from the first to the third century A.D.).

On the basis of conjectures and indirect evidences, one is led to think that the theoretical discussion at the beginning of the work comes to a great extent from Strato's work.[36]

The conclusions of historians lead then to attribute to Strato a disagreement with Aristotle on two important questions: the negation of the existence of void and the existence of two natural inclinations (one of the heavy bodies toward the centre of the Earth, the other of the light bodies in the opposite sense).[37]

1.3 After Aristotle in Greece and in Rome

After the School which continued Aristotle's Lyceum, two other philosophical Schools flourished in the Greek world: the Epicurean School and the Stoic School. Although both attended to the science of nature (they divided the science in three parts, ethics, physics and logic), this was not for improving the scientific investigation but rather for ensuring happiness to men. In fact, they thought that man, to be happy, ought to be free from anxiety and fear. For this, it was necessary to know the causes of the natural phenomena.

Generally speaking, one can say that Epicurus (341–270 B.C.) referred back to the doctrine of the Atomists, although criticizing their philosophy in some points.

A concise exposition of Epicurus' natural philosophy can be found in the **Letter to Erodotus** (which is a sort of syllabus dedicated to his disciple Herodotus, but also to others).[38]

Nor in this work neither in the other writings which have remained, Epicurus has dealt with the problems of motion from the point of view we are interested in.

---

[36]See: **A Source Book in Greek Science**, op. cit. pp. 248–255.

[37]The ascending tendency of air and fire could be explained by the movement downward of the heavier bodies.

[38]Epicurus says in the introduction to his little work: «For those, Herodotus, who can neither master all my physical doctrines nor digest my lengthier books On Nature, I have written a summary of the whole subject in enough detail to enable them to easily remember the most basic points, and thereby grasp these important and irrefutable principles entirely on their own, to whatever degree they take up the study of physics. Even those who have thoroughly learned the entire system must be able to summarize it, for an overall understanding is more often needed than a specific knowledge of details.» This English text is taken from Epicurus.info: E-Texts.

As regards the motion of atoms:

«Some get separated by great distances from each other. Other oscillate in one place whenever they happen to get entangled into a compound, or surrounded by a compound. It is the nature of both bodies and void which allows this oscillatory motion. For the bodies, being solid, rebound on collision to whatever distance their entanglement allows them, while the void offers no resistance in the intervening space. This may continue until at last the repeated shocks bring on the dissolution of the compound. There is no beginning to all these motions; the atoms and void are eternal.»[39]

Two centuries later, Epicurus found a very important follower in Rome: the poet Titus Lucretius Carus. Lucretius' poem **De Rerum Natura**, besides being a work of great poetic value is also a didactic work on Epicurus' philosophy.

Let us compare these verses with the excerpt of the **Letter to Herodotus** quoted above:

«For since they wander through the void, it must needs be that all the first-beginnings of things move on either by their own weight or sometimes by the blow of another. For when quickly, again and again, they have met and clashed together, it comes to pass that they leap asunder at once this way and that; for indeed it is not strange, since they are most hard with solid heavy bodies, and nothing bars them from behind.»[40]

As regards the study of natural phenomena, as we have already said, the Stoics took an attitude analogous to that of the Epicureans. The three thinkers who can be considered the founders of the Stoicism are Zeno of Citium (335–263 B.C.), Cleanthes of Asso (331–232 B.C.) and Chrisippus of Soli (280–207 B.C.).

The philosophical current of Stoicism lasted some centuries and usually is divided in three periods (the ancient Stoics, the medium Stoa, and lastly the Stoics of the imperial period of Rome. To the medium Stoa belongs Posidonius of Apameia (135–51 B.C.), who is considered the most important thinker of the first century B.C.

He was one of the most versatile authors of the Graeco-Roman world and represented, in a certain sense, the link between the Greek culture and that of Rome.

Even if no one of his works has remained, there exist testimonies that he wrote many books of scientific popularization[41] which were later on transferred in other works of Latin popularizers.

Maybe this is the cause of the disappearance of the (so to speak) originals in Greek.

---

[39]Ibidem.

[40]«Nam quoniam per inane vagantur, cuncta necessest.
aut gravitate sua ferri primordia rerum,
aut ictu forte alterius. Nam <cum> cita saepe
obvia conflixere, fit ut diversa repente
dissiliant; neque enim mirum, durissima quae sint
ponderibus solidis, neque quicquam a tergo ibus obstet.»
De Rerum Natura, II, 83-88. The English text is taken from **A Source of Greek Science**, op. cit. pp. 212–213.

[41]See for instance: **Posidonio—Testimonianze e Frammenti**—a cura di Emmanuele Vimercati—Bompiani, 2004.

It is universally held that the two most famous works of scientific popularization of the Roman world (**Naturales Quaestiones** of Seneca and **Naturalis
Historia** of Pliny the Elder) are to a great degree indebted to Posidonius' work.
For what regards original scientific investigations of Posidonius, historians mention a measurement of the terrestrial meridian and a measurement of the (apparent)
Sun's orbit.[42] The measure of the meridian, largely smaller than the renowned
measure of Eratostenes (275–195 B.C.), was considered the most authoritative
until the time of Columbus, being accepted even by Ptolemy.

All what we are going to tell regards the Greek science in general and its influence on the culture of the Graeco-Roman world but does not concern our personal
theme of interest. From this point of view, we must remark that, after Aristotle and
Strato, for centuries the dynamics of the mass point (just for using modern terminology) did not make any progress.

It is obviously true that in an interval of about three centuries there have been
Archimedes, Eratostenes and, at last, Hero, but their activity did not absolutely
concern dynamics.

Let us see how, at the beginning of the third century B.C., Pappus in the eighth
book of his **Mathematical Collection** gave a picture of the mechanics (which for
a mathematician essentially consisted in the theory of the centres of gravity and
the five simple machines):

> «The science of mechanics, my dear Hermodorus. is not merely useful for many impor-
> tant practical undertakings, but is justly esteemed by philosophers and is diligently pur-
> sued by all who are interested in mathematics, since it is fundamentally concerned with
> the doctrine of nature with special reference to the material composition of the elements
> in the cosmos. For it examines bodies at rest, their natural tendency, and their locomotion
> in general, not only assigning causes of natural motion, but devising means of impelling
> bodies to change their position, contrary to their natures, in a direction away from their
> natural places. In this the science of mechanics uses theorems suggested to it by a consid-
> eration of matter itself.
>
> Now the mechanicians of Hero's school tell us that the science of mechanics consists of a
> theoretical and a practical part.
>
> The theoretical part includes geometry, arithmetic, astronomy, and physics, while the
> practical part consists of metal-working, architecture, carpentry, painting, and the manual
> activities connected with these arts.
>
> One who has had instruction from boyhood in the aforesaid theoretical branches, and
> has attained skill in the practical arts mentioned, and possesses a quick intelligence, will
> be, they say, the ablest inventor of mechanical devices and the most competent master-
> builder. But since it is not generally possible for a person to master so many mathematical
> branches and at the same time to learn all the aforesaid arts, they advise a person who
> is desirous of engaging in mechanical work to make use of those special arts which he
> has mastered for the particular ends for which they are useful. The most important of the
> mechanical arts from the point of view of practical utility are the following.

---

[42]See the testimony of Cleomedes (maybe flourished in the imperial period before Ptolemy):
Cléomède—Théorie Élémentaire (De motu circulari corporum caelestium)—Texte présenté,
traduit et commenté par Richard Goulet—Paris, Librairie Philosophique J. Vrin, 1980.

1. The art of the *manganarii* known also, among the ancients, as mechanicians. With their machines they need only a small force to overcome the natural tendency of large weights and lift them to a height.
2. The art of the makers of engines of war, who are also called mechanicians. They design catapults to fling missiles of stone and iron and the like a considerable distance.
3. The art of the contrivers of machines, properly so-called. For example, they build water-lifting machines by which water is more easily raised from a great depth.
4. The art of those who contrive marvelous devices. They too are called mechanicians by the ancients. Sometimes they employ air pressure, as does Hero in his Pneumatica; sometimes ropes and cables to simulate the motions of living things, e.g., Hero in his works on Automata and *Balances;* and sometimes they use objects floating on water, e.g., Archimedes in his work On Floating Bodies, or water clocks, e.g., Hero in his treatise on that subject, which is evidently connected with the theory of the sun dial.
5. The art of the sphere makers, who are also considered mechanicians. They construct a model of the heavens [and operate it] with the help of the uniform circular motion of water.»[43]

Therefore, the mechanics, as a mathematical discipline, did not provide for the dynamics of "local motion" of bodies and for a long time one wondered why the Greeks did not make any move forwards the mathematization of physical phenomena (except those of astronomical nature).

S. Sambursky, among the many historians who have tried to give an answer to this question, says:

«If we set the scientific maturity of Greek astronomy over against the feebleness of Greek achievements in "terrestrial" physics, we cannot help enquiring into the reason for this enormous contrast. The answer is to be found mainly in the great simplicity of astronomy when compared with the physical phenomena all round about us.»[44]

For simplicity of astronomy, Sambursky meant that the sky directly gave the possibility of observing a phenomenon (the motion of celestial bodies) which periodically repeated itself, without no need of human intervention.

Since

«The essential thing in an experiment is the isolation of a certain phenomenon in its pure form, for the purpose of studying it systematically.»[45]

What was certainly absent in the Greek scientific world was an active consent to a systematic experimentation. Then, if we want to sum up, the dynamics of bodies in the sublunary world that the Greek science left to us is that of Aristotle, with the corrections of Strato we have already mentioned. Substantially, the sublunary world consisted of a "plenum" (the existence of the void was in any case excluded) where both "natural" and "violent" motion (i.e. under constriction) occurred. The bodies were divided in heavy and light, with different graduations depending on component elements. The heavy bodies, with natural motion,

---

[43]See: **A Source Book in Greek Science**, op.cit., pp. 183–184.

[44]S. Sambursky: **The Physical World of the Greeks**—London, Routledge and Kegan Paul, 1963—Chap. X.

[45]S. Sambursky, Ibidem.

fell towards the centre of the Earth following a rectilinear path; the light bodies, instead, moved upward in a straight line towards the moon's sphere.

The earth was the absolutely heavy element and the fire the absolutely light one, water and air were intermediate elements which had only a relative heaviness and a relative lightness.

Both the motions downward and those upward was accelerated motions; their velocity was increasing as they got close to their natural place.

As regards the cause of natural motions, Aristotle traced it to a kind of remote generator, but in the case of the fall of a body identified in the weight the direct cause, as in the lightness identified the cause of the ascent upward. Other things being equal, the velocity of motion was directly proportional to the weight of the body and in inverse proportion with the resistance of the medium (due to density). The time taken in the motion was directly proportional to the resistance of the medium and in inverse proportion with the weight of the body.

The violent motions occur when the bodies are obliged to leave their natural places by an (external) motive force, for example a projectile thrown up in the air. In this case, the velocity of the body V will be directly proportional to the motive force F and in inverse proportion with the resistance R it met with; in formula $V \propto \frac{F}{R}$.

As regards the persistence of the motion (for instance the case of a projectile which continues its motion, even after it is no more in contact with whom has thrown it), the explanation of Aristotle is that we have seen in the quoted excerpt of **Physics** (footnote 30).

He maintained that the external medium, the air in the example of the projectile, was the source of the continuity of the motion: the prime mover sets in motion also the air all around, and the air in turn activates the air contiguous to it which will push on the projectile and will activate the further layer of air released by the projectile and so on. But in the subsequent phases the motive force (because of the resistance of the medium due to density) decreases until vanishing; at that point the projectile falls down with natural motion. Therefore, the medium has the dual function of motive force and resistance.

As we shall see, the criticism of the medieval philosophers will address on this.

Bibliography

We quote here only books and papers not mentioned in the footnotes of the chapter.

Aristotle, *Meteorologica* (Translated by H. D. P. Lee—(Loeb classical Library) Harvard U. P. 1952)

Aristotle, *On the Heavens* (Translated by William Keith Chambers Guthrie—(Loeb Classical Library) Harvard U. P. 1939)

Aristotle, *Physics* (Translated by P.H. Wick Steed, F.M. Comford—(Loeb Classical Librarie) Harvard U. P. 1934)

A.H. Coxon, *The Fragments of Parmenides—A Critical Text with Introduction and Translation, the Ancient Testimonia and a Commentary* (Parmenides Publishing, Las Vegas, 2009)

L. Edelstein, I.G. Kidd (eds.) *Posidonius: The Fragments* (Cambridge Classical Texts and Commentaries, Cambridge U. P. 1999)

E.S. Forster, *Mechanics*, in *The Complete works of Aristotle.* The Revised Oxford Translation, ed. by J. Barnes (Princeton University Press, Princeton, 1995)

R. Geer (ed.), *Epicurus: Letters Principal Doctrines and Vatican Sayings* (Pearson, 1964)

W.K.C. Guthrie, *The Greek Philosophers—From Thales to Aristotle* (Harper & Row, New York, 1975)

W.E. Leonard, S.B. Smith (eds.), *Titus Lucretius Carus: De Rerum Natura–The Latin text of Lucretius* (University of Wisconsin press, 2008)

Lucius Annaeus Seneca, *Natural Questions* (Translated by Herry M. Hine, University of Chicago Press, 2010)

Plato, *Timaeus, Critias, Cleitophon, Menexenus, Epistles* (Translated by R. G. Bury—(Loeb classical Library) Harvard U. P, 1999)

Serres M, *The Birth of Physics.* (Clinamen Press Ltd, 2001)

W.H. Sthal, *Roman Science* (The University of Wisconsin Press, Madison, 1962)

The Poem of Empedocles: A Text and Translation with a Commentary by Brand Inwood (Phoenix Supplementary volumes) (University of Toronto Press, 2001)

Chapter 2
The Theories of Motion in the Middle Ages and in the Renaissance

The period considered in this chapter (about ten centuries long) is that which, in its final part, is close to the time of Galileo and then, also in the light of the theories on the importance of some medieval works, it is essential to make clear in what context Galileo worked. This context includes a complex of theoretical treatments, starting from the earlier reworking and revisions of Aristotle's work in the Dark Ages until the work of the immediate forerunners of Galileo, that we are tempted to define as "the medieval mechanics" as a whole.

At first sight it may seem daring and superficial to give this appellation to the cultural production of a period which includes also the Renaissance. As a justification we can say that the works of Leonard, Tartaglia or even Benedetti makes one think more to an "autumn of the Middle Ages" (in the sense that one has to do with criticisms "from the inside") than to the dawn of a "new science". Obviously, as we say over and over again, our conclusion only concerns dynamics.

2.1 Preliminary Remarks

The cultural revival of the Middle Ages is usually traced back to the great work of translation (in Latin) of the works of Greek philosophers (above all Aristotle) both through their Arabic translations and directly from Greek.

As we know, the works of Aristotle mainly dedicated to natural philosophy are five (**Physics**, **On the Heavens** (**De Caelo**), **Generation and Corruption**, **Metheorologia** and **The Soul**). The work of translation in Latin of the Aristotelian corpus began starting from Boetius in the VI century, as above said, translating both second hand from Arabic, and directly from the available Greek manuscripts.

Almost all the translations were performed in the XII century. The most important translators from Arabic were Gerard of Cremona in the XII century and Michael Scot in the XIII century. Robert Grosseteste and William of Moerbeke were, in the

© Springer International Publishing Switzerland 2016
D. Boccaletti, *Galileo and the Equations of Motion*,
DOI 10.1007/978-3-319-20134-4_2

XIII century, the most important translators from Greek. If we look at the list of the works of the natural philosophers of the Middle Ages, apparently we meet always with the same titles; in fact all of them write commentaries to the works of Aristotle (**Commentary on the Physics**, **Commentary on De Caelo**, etc.).

But these commentaries were not a mere comment on Aristotle, in the sense in which we use the term nowadays; they contained discussions and revisions, and new theories that were proposed for correcting or replacing those of Aristotle. In this way, everyone vehiculated his ideas through the commentaries on "the philosopher". The diffusion of the ideas contained in these works obviously occurred through the manuscripts and the lectures that the authors gave in the universities of that time, predominantly in the so called Faculties of the Arts (in the disciplines of the Quadrivium).

Therefore, by using a locution in fashion nowadays, the catchment area was quite restricted and this fact also explained why only a few of these works were printed and made known with the advent of printing.

The historians of science of the XIX century therefore found themselves unaware of their existence. We can add that the same thing had already occurred in the preceding centuries. In fact, it is symptomatic that the work of Bernandino Baldi (1553–1617) **Le Vite de' Matematici** (written, it seems, in the penultimate decade of the XVI century), which is the first attempt of history of mathematical sciences made in the modern age,[1] does not mention the protagonists of the Parisian school of XIV century (Buridan, etc.) nor the English mathematicians of the Merton College. To be precise, Bradwardine and Swineshead (Italianized in Brauardino and Suisseto), besides to be quoted in the **Vite**, are inserted with brief mentions also in the little work **Cronica de Matematici** (printed in Urbino in 1707), wherefrom it can be inferred that it is second hand information; in any case, there is no mention of writings concerning the mechanics.

This circumstance could cast doubt on the supposed "large dissemination" of the theories of the Parisian school in Italy, but we shall come back on this topic. Instead, the "rediscovery" of the medieval mechanics occurred much later starting from the manuscripts preserved in the libraries. The principal "explorer", author of numerous discoveries, was the French physicist (and historian of science) Pierre Duhem (1861–1917) who recasted in thousands of pages his studies on the found medieval manuscripts. As to the part which directly concerns the mechanics, we cite the two works **Les Origines de la Statique**—Paris, Hermann (two volumes, 1905–1906) and **Études sur Léonard de Vinci**—Paris, Hermann (three volumes, 1909–1913). The third volume of this last work has as subtitle **Les Précurseurs Parisiens de Galilée**. The historical thesis of Duhem, that we shall discuss later on, is already expressed in the title. We also cite the monumental work in ten volumes **Le Système du Monde**—Paris, Hermann, 1913–1959, with subtitle **Histoire des Doctrines cosmologiques de Platon à Copernic**, three volumes of which (7°, 8°, 9°) are dedicated to "La Physique parisienne au XIV siècle". The last five volumes of this work were published posthumous from 1954 to 1959.

---

[1]See Bernardino Baldi: **Le Vite de' Matematici**—Edizione annotata e commentata della parte medievale e rinascimentale a cura di Elio Nenci—Milano, Franco Angeli, 1998.

Coming back to the above assertion concerning the historians of science of the XIX century, the work of Ernst Mach **The Science of Mechanics** (first edition 1883)[2] which—leaving aside the epistemological theses of the author—represents the first and authoritative critical study on the history of mechanics, for what regards the dynamics substantially begins with Galileo, without any references to medieval problems. This structure of the text remained unchanged in all subsequent editions.

In fact, Mach's book had seven editions during the author's lifetime, therefore Mach had the time to follow the first part of the scientific production of Duhem. In the preface of the seventh edition (1912), Mach, after Emil Wohlwill and Giovanni Vailati, thanks Pierre Duhem for his critical observations and in the chapter regarding the statics cites **Les Origines de la Statique**, even if not completely agreeing with the author.

In the chapter on dynamics he quotes, instead, a historical and critical contribution of 1905 of Duhem regarding the accelerated motion (and Galileo).[3] Obviously, he could not have read the third volume of the **Études** published in 1913. Also in this quotation, Mach expressed a partially different opinion on the methodology of Galileo.

In the years between the two world wars, the historical research on the medieval science had a notewhorthy development and originated, astride the fifties of the last century, also a series of works addressed to an audience larger than that of the community of experts.[4]

Duhem's theses were in part accepted and in part refused, but always discussed with great respect for the work of the French scholar. Beyond doubt the most balanced (and nourished of profound culture and erudition) criticisms to Duhem came from the German scholar Anneliese Maier (1905–1971).[5] An exception, if it

---

[2]See Ernst Mach: **The Science of Mechanics: A Critical and Historical Account of its Development**—Translated by T. J. Mc Cormac—Open Court Classics 1988. (The first edition in German language was published in Prague in 1883).

[3]See P. Duhem: **De l'accélération produite par une force constante—Notes pour servir à l'histoire de la Dynamique**—Congrés international de philosophie (Geneva 1905), p. 859.

[4]One of the first works of this type was the **Histoire de la Mécanique** of René Dugas (Neuchâtel, 1950), which was published with a preface by Louis De Broglie and was translated in English some years later (the English translation was reprinted by Dover in 1988). To this, Dugas added **La mécanique au XVIIᵉ siècle—Dès antécédents scolastique à la pensée classique** (Neuchâtel, 1954).

[5]Unfortunately, Mayer's work is almost unknown out of the exclusive circle of the specialists who deal with the history of philosophy and medieval science. Maier took her degree in Germany and then, after having completed her studies at Zurich and Berlin, passed in Italy in 1936 to look for manuscripts of Leibnitz (entrusted with a mandate by the Prussian Academy of Sciences). Starting from 1937 she resided in Italy almost consecutively and there published the greater part of her writings. She wrote always in German and this happened when by that time the scientific literature regarding the subjects she dealt with was for the most part in English language. Her writings were almost all published (in German language) by the Edizioni di Storia e Letteratura—Roma— since the fifties of the last century. Only some decades later an English translation of a selection of her essays on the medieval sciences and Galileo was published: **On the Threshold of Exact Science—Selected Writings of Anneliese Maier on Late Medieval Natural Phylosophy**, edited and translated with an Introduction by Steven D. Sargent—University of Pensilvania Press, 1982. For these reasons, as we said, her name has not crossed the border of the specialistic research and therefore the results of her research are known only through the quotations of the specialists.

can be defined so, was constituted by Antonio Favaro (1847–1922), the author of the national edition of Galileo's works, who always considered himself invested of the "sacred mission" of fighting against those he called "the detractors of Galileo". This tendency to transform in an almost personal matter every criticism to the work of Galileo coming from scholars of various origins in some cases risked to make less admissible his opinion even when it was correctly and scientifically grounded.

Among the detractors of Galileo Favaro included also (and above all) Duhem and opposed his conclusion which credited to the "doctores Parisienses" the authorship of the results universally credited to Galileo.

Rightly, in our opinion, he could conclude «Then, we trust to be able to conclude that the opinion expressed by Duhem is at least susceptible to a revision, in which we would the great precept not to be forgiven: in the critical studies on the theory of the sciences it is necessary to forbear crediting to non-modern authors assertions which do not appear as direct conclusions from the preliminaries they enunciate, or preliminaries necessary for the conclusions at which they arrive».[6]

In other occasions he was less fair, in respect of Duhem.[7] He was also much aggressive towards Raffaello Caverni, who was the first to deal seriously with the results of Galileo concerning the mechanics. Caverni studied the mechanics of Galileo in the fourth volume of his work **History of the Experimental Method in Italy** (1891–1900).[8]

The negative criticism of Favaro was followed by the isolation of the work (in the meantime the author was dead before completing his work) which was substantially excluded, for about half a century, from the debate of Italian scholars and consequently, ignored abroad. Actually, Favaro's opinion on Caverni was practically shared by the most outstanding scholars of that time (Enriques, Marcolongo, Mieli, etc.).[9]

---

[6]A. Favaro: **Galileo Galilei e i "doctores parienses"**—Rendiconti della Reale Accademia dei Lincei—Classe di scienze morali, storiche e filologiche. Serie Quinta. Vol. XXVII (1918), pp 139–150.

[7]In 1921, Favaro, on reviewing a work of a French author regarding the Italian though of XVI century, takes the opportunity to address accusations to Duhem. With regard to the studies of Duhem, he says: «... for having an exact enough opinion of the important question of which such studies show only one side, and perhaps the less important, it will not be inopportune to remind its highest origins, even if for this is necessary to unveil the back stage.» In substance, Duhem's studies on the school of Parisian terminists and the consequent opinion on Galileo (only a continuator of theirs) would have been originated (through Cardinal Dechamps and abbot Mercier) by a directive of Pope Lion XIII for promoting the neo-scholasticism. See: A. Favaro—**Galileo Galilei in una rassegna del pensiero italiano nel corso del secolo decimosesto**—Archivio di storia della scienza, 2, 137–146 (1921).

[8]Raffaello Caverni: **Storia del Metodo Sperimentale in Italia** (six volumes)—Firenze, Stabilimento G. Civelli Editore (1891–1900). The work has had two anastatic reprints (Forni, Bologna 1970 and Johnson Reprint Corporation, New York/London, 1972).

[9]See: D. Boccaletti: **Raffaello Caverni and the society for the progress of the sciences: an independent priest criticized by the lay scientists**—Physis—vol. XLVIII (2011–2012) Nuova Serie—Fasc. 1–2.

2.2   The First Substantial Criticisms to Aristotelian Mechanics—Philoponus and Avempace

The Aristotelian theories concerning the motions in general and the fall of heavy bodies in particular were destined to last until the XVI century, but this did not prevented them from meeting with criticisms and oppositions already in the late Antiquity. In fact, the Greek world of the late Antiquity contributed, with an impressive quantity of commentaries to the works of Aristotle, to the development of the natural philosophy. A relevant criticism came from John Philoponus[10] (called the grammatist) who expressed his ideas in the VI century in the commentary to the **Physics** of Aristotle.

Obviously, we must not look in the commentary of Philoponus for elements which foreshadow the setting in of a new mechanics, but rather the elements of criticism to Aristotle, that is the enfeeblement of the Aristotelian tenets which began in this way to be subjected to a critical revision. A fundamental point, for instance, is this: Philoponus asserts, on the contrary of Aristotle, that the existence of void is possible and then he speaks of motion in the void as well. In the opinion of Philoponus, the fundamental and primary entity that determines the motion is the motive force.

If a body moves in the void, the motive force makes it to walk a certain space in a certain time. If instead the body moves in a certain medium, it meets with a certain resistance which is in direct proportion to the density of the medium and therefore one must add an additional time to the primary time (i.e. that taken to move in the void). Therefore, Philoponus rejects the Aristotelian theory which identifies in the ambient medium the "motor conjunctus" of a "projectum separatum".

By anticipating, in a certain sense, the theory of impetus he asserts that the motive force must be considered as the "motor conjunctus" that the "projector" has imparted to the "projectum" when throwing it. With regard to the fall of heavy bodies, Philoponus agrees with Aristotle that the heavier bodies fall with higher velocity. This happens also in the void:

«…. And if bodies possess a greater or a lesser downward tendency in and of themselves, clearly they will possess this difference in themselves even if they move through a void. The same space will consequently be traversed by the heavier body in shorter time and by the lighter body in longer time, even though the space be void. The result will be due not to greater or lesser interference with the motion but to the greater or lesser downward tendency, in proportion to the natural weight of the bodies in question…».[11]

The work of Philoponus remained unnoticed in the Latin West until the XVI century (Galileo knew a Latin translation published in 1535—quoted by him many

---

[10]John Philoponus (about 490–570) was a Byzantine philosopher of Greek language (Neoplatonist and also Christian) and also director of the School of Alexandria.

[11]See: **A Source Book in Greek Science** by M. R. Cohen and I.E. Drabkin, Harvard University Press—1966, p. 217.

times, but not in relation with the theory of impetus). Sentences, in a certain sense connected to those of Philoponus, can be found in the work of the Spanish Muslim Ibn-Badja, known to the Latin scholastics under the name of Avempace,[12] at a distance of six centuries.

In the opinion of Avempace, on the contrary of Aristotle, the medium is not essential for the natural motion with finite velocity since the velocity of motion is determined by the difference and not by the ratio, between the density of the body and that of the medium. Therefore, $V = F - R$, so that when $R = 0$, $V = F$ (F is the motive power measured by the specific gravity of the body which moves, R the resistance of the medium measured by its specific gravity, V the velocity).

It must be noted that Avempace did not say how the motion, in the magnitudes characterizing it, could be then really measured, at least in the exposition by Averroes who, in his turn, confuted him in his famous commentary to the **Physics** of Aristotle. It does not fall into our purpose to go on to explain the theory of Averroes etc. since this will bring us far from the subject in which we are interested, i.e. the theories which directly foreshadow the work of Galileo.[13]

2.3 The Medieval Kinematics

The concept of motion in the scholastic Philosophy deriving from Aristotle is broader than the simple reference to the change of position of a body in the space, with the relative attributions of velocity etc. as we are used to mean from Galileo on. According to the Scholastics, the motion was a transition from the potentiality to the actuality and vice versa and then regards any case in which one would appeal to the distinction between actual and potential. According to Aristotle

«.... Again, there is no such thing as motion over and above the things. It is always with respect to substance or to quantity or to quality or to place that what changes changes.»[14]

Here is also important to remind that the philosophy of Aristotle could not be completely accepted in its "original" version by the Christian world. In fact, according to Aristotle, the world existed ever since and therefore could not have been created.

Jointly with discussions, that we can call restricted or particular, about several points of Aristotle's works regarding the philosophy of nature, there was a fundamental "revision" of the Aristotelian philosophy to graft it in the Christian

---

[12]Ibn-Badja is the first philosopher famous among the Arabs of Spain—he was born in Saragossa at the end of the XI century and died in Fez in 1138.

[13]It is of a great interest, on the discussions regarding the Aristotelian physics in the Middle Ages, the long essay of E. A. Moody: **Galileo and Avempace—The dynamics of the leaning tower experiment**—published in two parts in the Journal of the History of Ideas Vol. 12 (1951. 163–193, 375–422). We shall refer to this essay in Chap. III.

[14]See: Aristotle: **Physics**, (III, 1–200 b 30).

philosophy. One arrived also at a clear distinction between theology and philosophy, but also with the clear statement that the philosophy was "ancilla theologiae".

The authors of this work of "Christianization" were essentially Albert the Great (Albertus Magnus) (1206–1280), who studied the commentators of Aristotle who preceded him starting from Avicenna, and St. Thomas Aquinas (1225–1274).

After them, with reference to this work, one will ever speak of "Thomistic synthesis", even if the "synthesis" was not entirely due to Thomas.

In the work of the natural philosophers of the Middle Ages, as on the other hand in the treatments of the motion in the Greek world, the description of the motion and the attributions of their causes are often interconnected, so in these cases it appears hard to disaggregate the kinematics from the dynamics, as in the modern textbooks of rational mechanics.

However, we shall try to deal separately with the two subjects since this will help us in the task of the subsequent comparison with the results obtained by Galileo. A simple case, i.e. of a work which is merely a treatise of knematics, is supplied by the **Liber de Motu** of Gerard of Brussels.

2.3.1 Gerard of Brussels and the Liber de Motu

The manuscript of the **Liber de Motu** was discovered by Duhem into the Latin fund of the French National Library and then briefly summarized in the third volume of his Études.[15] Subsequently, Eneström[16] told of it, and, finally, it was published (for the first time in 1956 and in final edition in 1984) by Marshall Clagett.[17]

We are still out of reliable information on the author, except for the name; the only certain date is a "terminus ante quem" indirectly fixed for the date of composition of the work (1260). Then the work surely goes back to the first half of the XIII century[18] and «it is perhaps the first, certainly one of the most important medieval works dedicated to kinematics» (E. Giusti).

The problem that Gerard deals with is that of the velocity of extended bodies in uniform rotary motions. Although Clagett has emphasized the influence of Euclid and Archimedes on Gerard, the work «to the best of our knowledge, is a completely original creation of the philosopher of Brussels» (Giusti). The **Liber**

---

[15]P. Duhem: **Études sur Léonard de Vinci**—Troisième série—1913, pp 292–295.

[16]G. Eneström: **Sur l'auteur d'un traité De Motu au quel Bradwardine a fait allusion en 1328**—Archivio di storia della scienza 2, 133–136, 1921.

[17]M. Clagett: **The Liber de Motu of Gerard of Brussels and the Origin of the Kinematics in the West**-Osiris 12, 73–175, 1956. M. Clagett: **Archimedes in the Middle Ages**, vol. V—Madison Wisc., 1984.

[18]For this and a thorough discussion of the **Liber de Motu**, see E. Giusti: **Alle origini della cinematica medievale: il Liber de Motu di Gherardo da Bruxelles** (Bollettino di Storia delle Scienze Matematiche—vol. XVI, 199–240, 1996).

is divided into three parts (books). The first book deals with the rotation of segments, the second with the rotation of plane figures, the third with the rotation of solid bodies. We refer the reader to the quoted paper of E. Giusti for an exhaustive examination of the cases dealt with by Gerard.

What is important for us is to point out that the work of Gerard arrives at about one thousand four hundred years after the death of Archimedes (212 B. C.), that is, of the last author who had dealt with the uniform motion.

Moreover, we consider important, also prior to a comparison with the kinematics of the Merton College, to point out what separate the definition (really, as we know, it cannot properly be called a definition) of velocity given by Gerard from that of Aristotle. Gerard says:

«Proportio motuum punctorum est tamquam linearum in eodem tempore descriptarum».[19]

The translation quoted in footnote 19 is that due to Clagett who, on the other hand, agrees with Giusti in translatig the Latin "motus" by "velocity". What is the essential point? Aristotle says in the **Phisics** (VI 2, 232 a):

«…. it necessarily follows that the quicker of two things traverses a greater magnitude in an equal time, an equal magnitude in less time, and a greater magnitude in less time, in conformity with the definition sometimes given of "the quicker"….».

That is, as we have many times stressed, in the Greek world one was bound to use the proportion which were constructed by ratios of homogeneous magnitudes and these, in the case we are interested in, were distances and times: a ratio between two distances was compared with the ratio between two times. Gerard, for the first time, speaks of velocity as a magnitude by itself, that is, the velocities enter directly in the proportion and are no more indirectly comparable starting from a proportion which does not contain them. This really represents a novelty, even if it will be necessary to wait for more than four centuries to arrive at the concept expressed by the ratio $v = \frac{s}{t}$. Doubtless, it is an advance towards a true definition of velocity, even if overvalued by somebody.[20]

2.3.2 The Kinematics at Merton College

We have seen above (cf. footnote 16) that one of the first studies on Gerard of Brussels refers to the citation and subsequent discussion made by Thomas Bradwardine in his

---

[19]See the final edition of Clagett quoted in footnote 17, p. 64: «The proportion between the velocities of points is the same as that between the lines described in the same time».

[20]See: J. Mazur: **Zeno's Paradox**, Dutton, 2007. J. Mazur asserts that with Gerard one has to do, for the first time, with velocities considered as magnitudes and such an approach marks a turning point in the direction of the modern concept of istantaneous velocity. But neither Clagett, nor Giusti and not even Souffrin (who had dealt thoroughly with the concept of velocity) had made an assessment of this kind of the quoted passage of Gerard. The paper of Souffrin we refer to is the already quoted **Sur l'histoire du concept de vitesse d'Aristote à Galilée**—Medioevo—Rivista di Storia della Filosofia medievale—vol. XXIX (2004) pp 99–133.

Tractatus de Proportionibus (1328).[21] Bradwardine is doubtless the first moving spirit of the group that had developed at the Merton College of the University of Oxford and was working about the half of the XIV century. The components of this group were, besides Bradwardine (about 1300–1349), William Heytesbury (1313–1373), Richard Swineshead (?–1355) and John Dumbleton (?–1349).

The most important works from our point of view (i.e. those which have something to do with mechanics) produced by this group are the already mentioned **Tractatus de Proportionibus** (1328), the **Regule solvendi Sophysmata**[22] of Heytesbury and the **Liber Calculationum**[23] of Swineshead which won for his author the nickname "the calculator".

The biographical data regarding these four authors are scarce and doubtful and there are some doubts also about the other works ascribed to them. For a more detailed discussion we refer the reader to the classical work of Marshall Clagett[24] containing also an extended bibliography (even if it stops at 1959).

The important thing for us is to see what progress had been made by the philosophers of Merton College in the study of mechanics and what footholds they had gained wherefrom the mechanics could have started again. Of course, we must always keep in mind that the results obtained by these authors should not be appraised from the modern (post-Galilean) point of view, i.e. that of estimating how much of "scientific" they contain. The interests of the philosophers of Merton College ranged from theology to logic and mathematics, and kinematics was not the predominant interest. Rather, if we can say so, it came as a particular application of the philosophical problem of how the qualities (or other forms) increase in intensity. In the scholastic terminology it was the problem "de intentione et remissione formarum".

The philosophers of Merton College, among the qualitative variations, considered also the problems of motion in the space and then the variations of velocity. Marshall Clagett summarizes in this way the contributions of the Mertonians to the development of the Mechanics:

1. A clear distinction between *dynamics* and *kinematics*, expressed as distinction between the *causes* and the space-time *effects*.
2. A new approach to the swiftness or velocity, within the ambit of which the idea of an instantaneous velocity was considered, perhaps for the first time,, and the idea of "function" was specified.

---

[21]The editions in modern languages that one can look up are:

1. H. Lamar Crosby: **Thomas Bradwardine. His Tractatus de Proportionibus. Its Significance for the Development of Mathematical Physics** (Madison, Wis., 1955).
2. Thomas Bradwardine: **Traité des Rapports entre Les Rapidités dans Les Mouvements**—suivi de Nicole Oresme: **Sur les Rapports**—Introduction, traduction, et commentaires de Sabine Rommevaux-Paris-Les Belles Lettres, 2010.

[22]To date complete translations in a modern language do not exist. For excerpts, one can see Clagett, op. cit., second part and Curtis A. Wilson: **William Heytesbury-Medieval Logic and The Rise of Mathematical Physics** (Madison, Wis., 1956).

[23]For a significant excerpt (in a modern language), see Clagett, op. cit. Chap. 5.3.

[24]Marshall Clagett: **The Science of Mechanics in the Middle Ages**—University of Wisconsin Press, 1959.

3. The definition of the uniformly accelerated motion considered as that motion in which equal increases of velocity are obtained in equal intervals of time.
4. The formulation and the demonstration of the fundamental kinematic theorem which establishes the equality, with respect to the space covered in a given time, of a uniformly accelerated motion and of a uniform motion whose velocity is equal to the velocity of the accelerated motion at the half of the time of acceleration.

With regard to the first point, the distinction is already clearly expressed in the Prologue (presently ascribed to a scribe) of Bradwardine's treatise where the contents of the four chapters of which it is composed were listed:

«... The third chapter makes clear the meaning of the ratio between the velocities of motion in comparison with the things moved and of the movers ... The fourth chapter investigates the ratio between the velocities of motion in comparison with the quantities of moveable and of the space covered ...».

As one can see, also Bradwardine, like the other Mertonians, puts the dynamics before the kinematics, contrary to the modern use. For the second and the third point, the most suitable reference is in a passage of the **Regule solvendi Sophysmata**[25] of Heytesbury where the definitions of uniform velocity, uniform acceleration and instantaneous velocity appear (the non-uniform motion was named "difformis" and that with constant acceleration "uniformiter difformis"). The definition of uniform motion is further specified (that is, the velocity must be the same in any fraction of time however small) in a writing ascribed to Swineshead.[26]

We quote here some excerpts:

«Of local motions, then, that motion is called uniform in which an equal distance is continuously traversed with equal velocity in an equal part of time. Non-uniform motion can, on the other hand, be varied in an infinite number of ways, both with respect to the magnitude, and with respect to the time.» ... «In non-uniform motion, however, the velocity at any given instant will be measured (attendetur) by the path which would be described by the most rapidly moving point if, in a period of time, it were moved uniformly at the same degree of velocity (uniformiter illo gradu velocitatis) with which it is moved in that given instant, whatever [instant] be assigned.» ... «With regard to the acceleration (intensio) and deceleration (remissio) of local motion, however, it is to be noted that there are two ways in which a motion may be accelerated or decelerated: namely, uniformly, or non-uniformly. For any motion whatever is uniformly accelerated (uniformiter intenditur) if, in each of any equal parts of the time whatsoever, it acquires an equal increment (latitudo) of velocity. And such a motion is uniformly decelerated if, in each of any equal parts of the time, it loses an equal increment of velocity.».

And finally Richard Swineshead was careful to specify that «uniform velocity is to be defined by the traversal of an equal distance in every (omni) equal period of time.».

---

[25]See Clagett, op. cit. ibidem and also **A Source Book in Medieval Science** (edited by Edward Grant-Harvard University Press, 1974) p. 238.

[26]See Clagett, op. cit., ibidem.

Let us pass now to that named fourth point by Clagett. In this case we must dwell longer since we have to do with the most important result obtained by the Mertonians: the so-called theorem of uniform acceleration or of the mean velocity. Substantially, one has to evaluate the space covered in a uniformly accelerated motion, when starting both from rest and from a point reached with a certain velocity. If this velocity is indicated by v_0 and the final by v_f and a indicates the constant acceleration for passing from v_0 to v_f, we should write for the covered space $s = v_o t + \frac{1}{2}at^2$, where t is the time taken for passing, with uniformly accelerated motion, from initial velocity v_0 to final velocity v_f. We also know that it must be $v_f = at$.

The Mertonians' theorem maintains that the uniformly accelerated motion is equivalent to a uniform motion with a velocity equal to that of the accelerated motion at half of its path. We can conflate all in the modern formula $s = \left[v_o + \left(\frac{v_f - v_0}{2} \right) \right] t$. In the case where the accelerated motion starts from rest ($v_0 = 0$), we shall have $s = \frac{1}{2} v_f \cdot t = \frac{1}{2}at^2$ (having taken into account in the last equality that $v_f = at$). In the case of $v_0 \neq 0$, being $v_f = at + v_o$, we shall have the known formula quoted above: $s = v_o t + \frac{1}{2}at^2$.

Let us see now how Heytesbury expressed what we have said above in his **Regule** (almost certainly it is the oldest enunciation of the theorem):

«From the foregoing it follows that when any mobile body is uniformly accelerated from rest to some given degree [of velocity], it will in that time traverse one-half the distance that it would traverse if, in that same time, it were moved uniformly at the degree [of velocity] terminating that latitude. For that motion, as a whole, will correspond to the mean degree of that latitude, which is precisely one-half that degree which is its terminal velocity.

It also follows in the same way that when any moving body is uniformly accelerated from some degree [of velocity] (taken exclusively) to another degree inclusively or exclusively, it will traverse more than one-half the distance which it would traverse with a uniform motion, in an equal time, at the degree [of velocity] at which it arrives in the accelerated motion. For that whole motion will correspond to its mean degree [of velocity], which is greater than one-half of the degree [of velocity] terminating the latitude to be acquired; for although a non-uniform motion of this kind will likewise correspond to its mean degree [of velocity], nevertheless the motion as a whole will be as fast, categorematically, as some uniform motion according to some degree [of velocity] contained in this latitude being acquired, and, likewise, it will be as slow.».[27]

Several demonstrations (Probationes) have been given of the above theorem. One can read the text of three of these demonstrations (authors: Heytesbury, Swineshead and Dumbleton) in Chap. 5 of Clagett. In the demonstration of Heytesbury, after the "theorem of the mean velocity", also the so-called "law of the distances" is demonstrated, in which it is maintained that a body which moves with uniformly accelerated motion starting from rest covers, in the second half of the time, a path threefold greater than in the first.

This law will be subsequently generalized by Oresme and we shall find it demonstrated by Galileo in general form.

---

[27]See Clagett, op. cit., ibidem and **A Source Book in Medieval Science**, op. cit. pp 239–240.

2.3.3 The Kinematics of the Parisian School

The results obtained by the philosophers of Merton College spread fast in the universities of that time until being reworked at the University of Paris by the local masters, the so-called *doctores parisienses*. Here too, we can speak of three authors who, besides belonging as masters of logic to the movement of the *terminists*, have dealt with the mechanics. They were Jean Buridan, latinized Buridanus (about 1295–about 1358), Nicole Oresme (about 1220–1382), and Albert of Saxony (about 1316–1390). We shall deal with the theory of Buridan (the impetus theory) further on, having chosen to put the dynamics after the kinematics.

The most significant and, say, innovative results in the field of kinematics are due to Oresme.[28]

Even if the historians of science have pointed out that Oresme had forerunners in the introduction of a geometric method in the study of kinematics,[29] it is usual to consider him the founder of this method since the clearest and rigorous formulation of the method is due to him. Granted that all work (both in Latin and in French) of Oresme has been left handwritten until the present time and with titles often assigned later by copysts, the two works that are of interest for the subject we are dealing with bear the title **Tractatus de configurationibus qualitatum et motuum**[30] (1350), and **De proportionibus proportionum**,[31] respectively.

The basic idea of the **Tractatus de configurationibus** is the following: the quantity of a quality can be represented by a geometric figure. The old problem *de remissione et retentione formarum* is geometrized.

The extension of quality (for instance, a time interval in the case of a motion) is represented by a part of a horizontal line while the qualitative intensity (for instance, the velocities at the different instants of the above interval of time) are represented by vertical segments perpendicular to the line of extension (see Fig. 2.1).

In the case of a uniform motion (i.e. with constant velocity) it is clear that the segments which represent the intensities will be all equal.

If we represent with segment AB a given interval of time and with AC and BC the equal velocities at the initial and final instants, we shall have a rectangle ABCD where the segments representing the velocities at the different instants will be contained, i.e. the figure represents the whole distribution of the intensities in

---

[28]For the overall work of Oresme and the scanty biographical data on him we refer to the article of Clagett in Charles Coulston Gillispie—**Dictionary of Scientific Biography**—Scribners, New York 1970—vol. 9, pp 223–230.

[29]See Clagett, op. cit., ibidem and S. Rommevaux, op. cit. in footnote 21, pp LXII–LXVI.

[30]See **Nicole Oresme and the Medieval Geometry of Qualities and Motions: A treatise on the Uniformity and Difformity of Intensities known as Tractatus de configurationibus Qualitatum et Motuum**—edited and translated by Marshall Clagett—Madison, Wisc., 1968.

[31]See Nicole Oresme: **De proportionibus Proportionum—ad Pauca Respicientes** (ed. E. Grant)—Madison, Wisc., 1966. See also the edition in French language quoted in footnote 21.

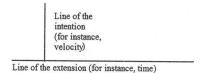

Line of the
intention
(for instance,
velocity)

Line of the extension (for instance, time)

Fig. 2.1

the quality, that is, the quantity of the quality. In the case of motion, it represents the whole space covered in the given interval of time (Fig. 2.2).

In the case of a uniformly accelerated motion (uniformly difform), since the velocity increases at a uniform rate, the relevant figure will be a right-angle triangle if the motion starts from rest (see Fig. 2.3). At this point, the (geometric) demonstration of the Mertonian theorem results quite immediate.

If ABC is the right-triangle representing the uniformly accelerated motion, it is immediate to control that rectangle ABFD, which represents the uniform motion with velocity equal to the velocity of the accelerated motion at half of the taken time (segment EG), has the same area (Fig. 2.4).

Therefore, the space covered in the same time is the same. It is obvious that a mirror image of Fig. 2.4 (which represents a uniformly accelerated motion) represents a uniformly slow motion.

We shall come back later on the use of Fig. 2.4 for representing the fall of heavy bodies. One must remark that Oresme's treatment of the kinematics always remained at an "abstract" level, that is, there has not been on his part any attempt of application to motions existing in nature.

The aforementioned motion of fall of the heavy bodies was interpreted by Oresme by making reference, although not completely in agreement, to the Buridan's theory and then as due to a continuous accumulation of impetus.

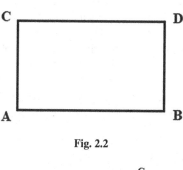

Fig. 2.2

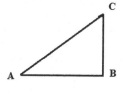

Fig. 2.3

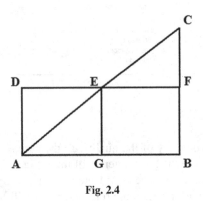

Fig. 2.4

2.4 The Medieval Dynamics

2.4.1 Bradwardine's Dynamics

As we have seen, Bradwardine separated the description of the motion (the kinematics, for us) from the study of the relation between the velocities of the motion and the magnitudes correlated to their causes. Substantially, in his work **Tractatus de proportionibus**, he takes again the question starting from the formulation of Philoponus, made known in those years by the translation from Arabic of the commentaries of Averroes to the works of Aristotle.

Bradwardine refuses a simple arithmetic proportionality of the type $V \propto (F - R)$, like that which can be inferred from the criticism of Philoponus to the theory of Aristotle (expressible through the relations $V \propto \frac{F}{R}$). The last becomes clearly meaningless when both $F = R$ (no motion, but V results different from zero), and $F < R$. Bradwardine's trouble was that of finding a kind of proportionality which could preserve a relation of proportionality between F and R but in the meantime could give vanishing velocities in the case $F = R$.

The solution was that «the proportion of the velocities in the motion follows the proportion of the power of the mover to the power of the thing moved». Said otherwise: the velocity increases arithmetically in correspondence of the geometric increase of the ratio of the force to the resistance. In formula, the relation can be understood, when varying F and R, as

$$F_2/R_2 = (F_1/R_1)^n, \quad \text{where } F_1/R_1 > 1 \text{ and } n = V_2/V_1.$$

In modern terms, one could express the relation in exponential form as

$$V = \log_a(F/R), \quad \text{where } a = F_1/R_1$$

Obviously, if $F = R$ one has $\log_a 1 = 0$ and then $V = 0$ for any value of a.[32] This result can be obtained anyway also in the rule of Philoponus $\propto (F - R)$.

2.4.2 Dynamics at the Parisian School and the Impetus—Theory

Bradwardine's law was largely accepted, except for some partial disagreement as we shall see later, until the beginning of the XVI century. Not even the *doctores parisienses* made an exception: Nicole Oresme and Albert of Saxony dealt with it in their works. Also their master, Buridan, was under the influence of the work of Bradwardine but was noticeable, above all, because of a new theory, generated by a radical criticism to the Aristotelian explanation of the motion of projectiles, gone down in history as the "impetus-theory".

According to Anneliese Maier[33] many people, starting from Duhem, have been interested in the impetus-theory imagining in it an anticipation of the inertia principle (Duhem was particularly insistent on this![34]).

Always quoting the opinion of Maier, «there is no doubt that it prepared the way for the law of inertia. The theory of impetus occupies an important and enduring place in the history of natural philosophy and physics as an independent stage of development between Aristotelianism and classical mechanics.[35]».

Obviously, an "exact analogous" of the inertia principle is beyond dispute. In fact, as we shall see, it was assumed that the uniform motion was due to a particular motive force (the impetus), whereas in the classical mechanics the uniform motion, like the rest, occurs in the absence of applied forces. More, since the universe was considered finite, the (rectilinear) uniform motion cannot last to infinity; only the circular motion of celestial bodies could last eternally. The only common point is: the motion remains uniform until an external force does not intervene.

The statements of Buridan (see further on) seem to confirm this, if isolated from the context. Anyway, the difficulty of interpreting the writings of those who

---

[32]This interpretation has been suggested by Anneliese Maier in **Die Vorläufer Galileis im 14 Jahrhundert**, Roma 1949, p. 92. We refer the reader to Clagett's work (op. cit. Chap. 7) and to the introduction of Sabine Rommevaux to the already quoted (see footnote 21) translation of Bradwardine's treatise for further widening. Our very swift and schematic synthesis has only the aim of recording which was the conclusion achieved in the criticisms and corrections to the Aristotelian mechanics in the ambit of Mertonians. Swineshead and Dumbleton further elaborated into details the theory of Bradwardine (see Clagett, op. cit. Chap. 7).

[33]See the essay: **The Significance of the theory of Impetus for Scholastic Natural Phylosophy** (in the volume **On the Threshold of Exact Science—Selected Writings of Anneliese Maier on Late Medieval Natural Phylosophy**—University of Pennsylvania Press, 1982)—The first edition of this essay (in German) is of 1955.

[34]Pierre Duhem: **Études sur Léonard de Vinci—Troisième série—Les Précurseurs parisiens de Galilée**—Paris-Hermann, 1913.

[35]In the essay quoted in footnote 33, p. 77.

have theorized on the impetus always lies in the used terminology, which for us is of complex translation and interpretation and therefore is not reproducible through the today's terms which always refer to quite precise definitions. In substance, the impetus is not a force, nor a form of energy, nor a momentum in the modern sense of the term. It shares something with these physical magnitudes but cannot be identified with anyone of them.

At the bottom of the impetus-theory there is always the Aristotelian principle "omne quod movetur ab aliquo movetur", therefore any motion entails the existence of a "virtus" or "vis motrix" as a cause.

The motion continues until the motive force exists and ceases when this vanishes. Of course these limitations immediately pose a number of questions. For instance, what is the cause of the motion of an arrow or a stone in-flight? The force impressed to a body is gradually attenuated by the medium and then the motion will vanish after a certain time because of the resistance met with. Necessarily, the velocity of a body will be directly proportional to the impressed force and in inverse proportion to the resistance. If the resistance remains constant the velocity depends only on the magnitude of the motive force.

Already Albertus Magnus (1206–1280) a century before the diffusion of the impetus-theory, expressed this concept: «**Omnis motus provenit ex virtutis moventis victoria super mobile, et cum illa virtus movet, oportet potentiam passivam eius, quod movetur, sibi esse proportionalem.**».[36]

Obviously, in this way, the velocity is proportional to the force, whereas in the classical mechanics the acceleration is proportional to the force.

The first to formulate the impetus-theory was the Italian Francesco di Marchia[37] (in a series of lectures given in Paris in the academic year 1319–1320). Afterwards, the man who became the principal supporter of the impetus-theory was Jean Buridan (1300– 1358),latinized in Buridanus, who exposed it in his commentary of the Aristotle's **Physics**.

Unlike of the structure essentially philosophical-theological of the work of di Marchia, Buridan refers to experience. Anyway, for both, the theory is as follows: in the instant in which the projector detaches himself from the projectile, this receives a secondary motive force (*impetus, vis impressa, vis derelicta* according to di Marchia) which is the cause of the subsequent motion. Let us see how Buridan enunciates his ideas on the fall of heavy bodies and their falling velocity.

In Question XII (whether natural motion ought to be swifter in the end than the beginning) in the commentary to book II of Aristotle's **De Caelo**:

«... one must imagine that a heavy body not only acquires motion unto itself from its principal mover, i.e., its gravity, but that it also acquires unto itself a certain impetus with that motion. This impetus has the power of moving the heavy body in conjunction with

---

[36]Albertus Magnus, **Physica** VIII, tract. II, cap. 6 (**Opera**, ed. Borgnet, Paris 1890) «Every motion is originated by the victory of the motive force over the moveable and when that force operates, it is necessary that the passive potentiality of the thing moved is proportional to it.»

[37]Di Marchia was a Franciscan and follower of Duns Scoto (the *doctor subtilis*).

the permanent natural gravity. And because that impetus is acquired in common with motion, hence the swifter the motion is, the greater and stronger the impetus is. So, therefore, from the beginning the heavy body is moved by its natural gravity only; hence it is moved slowly. Afterwards it is moved by that same gravity and by the impetus acquired at the same time; consequently, it is moved more swiftly. And because the movement becomes swifter, therefore the impetus also becomes greater and stronger, and thus the heavy body is moved by its natural gravity and by that greater impetus simultaneously, and so it will again be moved faster; and thus it will always and continually be accelerated to the end. And just as the impetus is acquired in common with motion, so it is decreased or becomes deficient in common with the decrease and deficiency of the motion. And you have an experiment [to support this position]: If you cause a large and very heavy smith's mill [i.e., a wheel] to rotate and you then cease to move it, it will still move a while longer by this impetus it has acquired. Nay, you cannot immediately bring it to rest, but on account of the resistance from the gravity of the mill, the impetus would be continually diminished until the mill would cease to move. And if the mill would last forever without some diminution or alteration of it, and there were no resistance corrupting the impetus, perhaps the mill would be moved perpetually by that impetus.»[38]

As we can see, in the final part of the quoted excerpt, Buridan also says how the impetus goes out in the case of a circular motion.

And, as regards the violent motion, i.e. of the projectiles, in Question XIII (whether the projectiles move swifter at half-way than at the beginning or at the end):

«And you see that the projector who moves the projectile is for some time tied with the projectile, continuously pushing the projectile before its ejection; like this, a man who casts a stone moves his hand with the stone, and also in shooting an arrow the string moves for some time with the arrow pushing it; and same also is for the sling which throws the stone, or for the machines which throw much bigger stones. And then, as long as the projector pushes the projectile which exists together with him, the motion is slower at the beginning, since only then the extrinsic mover moves the stone or the arrow; but during the movement an impulse (impetus) is acquired continuously, which combined with the extrinsic mover moves the stone or the arrow, which for this move swifter. But after the ejection from the projector, the projector does not move anymore, but only the acquired impulse (does move), as we shall see elsewhere; and that impulse, because of the resistance of the medium, weakens continuously, for which the motion gets continuously slower. And therefore, one must understand that the violent motions, i.e. those of projectiles, are swifter at the beginning than half-way or at the end, of course excluding that part of motion when the projector is together with the projectile; in fact, considering the remaining motion as a whole, the greatest velocity occurs at the beginning. And in this way the authoritative opinions of Aristotle and of the others must be reconciled. It is true that on this regard I have a doubt, since some say that the arrow thrown by the bow would be more perforating at a distance of twenty foot than at a distance of two foot, and therefore after the ejection from the bow the greatest velocity would be not yet at the beginning. And I have not experienced this, therefore I don't know if it is true; but if it were true, some say that the impetus is not immediately generated by motion, but continuously as a consequence of motion; and then it is not completely generated at the ejection from

---

[38]Iohannis Buridani: **Questiones super Libris quattuor de Caelo et Mundo**—edited by E. A. Moody—Cambridge, Massachusset, 1942, Book II, question 12, p. 180. The translation of this excerpt (by M. Clagett) is taken from **A Source Book in Medieval Science**, op. cit. p. 282.

the bow, but is accomplished in some time, as the rarefaction and the evaporation follow the heating, but not perfectly at once; indeed, once the heating has ceased, for the water is removed from the fire, yet for some time rarefaction and evaporation are seen to continue. And thus it is clear.»[39]

Here the theory of Aristotle is clearly refuted, which maintained that the mover of the projectile (after it has left the projector) is the surrounding air which receives by the projector an impulse which is transmitted from layer to layer, until its exhaustion. According to Buridan, instead, after the projectile has left the projector, the only mover is the acquired impetus.

2.5 The Diffusion in Italy of the Ideas of Mertonians and of the Parisian Masters

In the Middle Ages, the "innovative" ideas in the scientific field were chiefly disseminated by the masters of the universities in their transfers from one place to another. Particularly active was the switch of teachers and students between Italy and France.

Among the Italians who enjoy a great reputation both as scholars and as critics of the new theories the Parmesan Biagio Pelacani (about 1359–1416), also called Biagio da Parma[40] must be undoubtedly mentioned. We have to do with a typical figure of an intellectual of that time, though at the greatest level of interdisciplinarity (physician, philosopher, mathematician and also astrologer): he has both studied and taught in Paris, in addition to Padua, Pavia, Bologna and Florence. As astrological consultant of princes and lords of that time he deserved from his contemporaries the cognomen of *doctor diabolicus*.

Because of this activity of him, he was up before the bishop of Pavia who ordered him to reprocess some of his theses (by luck of Biagio, the Counter-Reformation was yet to come).

We have mentioned this aspect of Biagio's activity since «… his astrological doctrine, if on the one hand assured him great honours and successes at his time (because of his fame of infallibility) on the other hand discredited him in the eyes of the positivist scholars of the ninenteenth century, though they exalted his writings of optics, statics and astronomy.»[41]

---

[39]**Questiones super Libris quattuor de Caelo et Mundo**, op. cit. Book II, question 13, pp 183–184. (Our translation).

[40]On Biagio Pelacani, besides Clagett op. cit., see also the articles of F. Barocelli and G. Federici Vescovini in **Filosofia Scienza e Astrologia nel trecento europeo**, a cura di Graziella Federici Vescovini e Francesco Barocelli—Il Poligrafo, Padova, 1992–.

[41]Ibidem. G. Federici Vescovini is the most authoritative scholar of the work of Biagio; she has dealt with it for several occasions, in books and articles.

Biagio manifested a great interest both in the new mechanics of the Mertonians and in the phisics of Buridan. Limiting ourselves, according to our choice, to these subjects, we mention two works

1. **Quaestiones super tractatum de proportionibus Tomae Bradwardini.**
2. **Quaestiones de latitudinibus formarum.**
   The first of these works has had two different draftings (Florence 1388–1389 and Pavia 1389–1407). While in the first he accepts Bradwardine's rule of motion, in the second he challenged it for both mathematical and physical reasons.

In the **Quaestiones de latitudinibus formarum**, Biagio geometrically demonstrates the Mertonian theorem of the mean velocity in a way very similar to the demonstration of Heytesbury; but he follows Oresme by representing the lines of intensity by vertical instead of horizontal lines.

Besides Biagio Pelacani, the historians mention the presence in the universities, particularly of Padua and Pavia, of other philosophers interested in the doctrines both of the Mertonians and of the Parisian School. John of Casale and Franciscus of Ferrara flourished in the middle of the XIV century.

As regards John, he was supposed to have anticipated Oresme in the introduction of the use of the geometrical method in the study of the motions. Other names often mentioned are those of Angel of Fossanbruno and Jacopo of Forlì.

In short, one can anyway conclude that both Biagio and the others did not add elements of novelty to the theories of Mertonians and of Parisians, such as to modify their structure.

As a marginal note, we mention that the contributions of these philosophers in the field of mechanics (as of the other disciplines) circulated through copies of the manuscripts in the world of the universities (then in a restricted and exclusive community) and therefore were subject to getting out of the circuit following the disappearance of the authors.

Of the works we have mentioned before, only the **Liber Calculationum**[42] of Swineshead and the **Questiones de Latitudininibus** of Biagio Pelacani remained in circulation and intercepted the development of the art of printing, consequently having several editions in the last two decades of the XV centuries and the two first decades of the XVI. This circumstance, in our opinion, must be considered and accurately valued in the particular cases when one wants to credit forerunners for scientific results obtained in the subsequent centuries.

---

[42]With regard of what could have been the influence of this work in the Study of Padua, see the interesting paper of Christopher J. T. Lewis: **The Fortunes of Richard Swineshead in the time of Galileo**—Annals of Science, 33 (1976), 561–584.

2.6 The Theory of Motion in the XVI Century

Having as primary aim that of looking at the birth and the development of the theories of motion, starting from the Greek antiquity for arriving at Galileo, one is disappointed in seeing what happens in the XVI century. In fact, from this point of view the XVI century gives an image of itself various enough, if not confused. We mean that, in the case of that part of the mechanics regarding the theory of motion, one is not able to notice a continuous flourishing as in the literature, in the figurative arts, in the architecture and also in the technology (all what, i.e., delivered the idea of Renaissance in the Burckhardt's meaning), and not discerns an underlying theme which characterizes a tendency towards something of definite.

Maybe this impression comes from the fact that, so to say, one wants to begin from the end, that is, to single out the path (a mix of elaborations and results) which leads to the Galileian enunciations.

As it is known, starting from the last years of XV century (in parallel with the development of the art of printing), in Italy, and later on in the rest of Europe, several translations and commentaries of the works of the great Greek mathematicians (Euclid, Achimedes, Hero, etc.) have been published. Translations start to appear not only in Latin, but also in vernacular Italian and in other national languages.

The development of the technology motivates its professionals (who are not able to read the Latin—we remember that even Leonard styles himself as "omo sanza lettere") to demand mathematical texts, also those regarding machines.

Then, the diffused works are **On the Equilibrium of plane Figures** and **On floating Bodies** of Archimedes, the **Elements** of Euclid, and the **Mechanics** (also called **Questions of Mechanics**), work at that time attributed to Aristotle, but actually still of controversial attribution.

Also the writings **De Ponderibus** of the medieval mathematician Jordanus of Nemore (perhaps one had to do with several persons under this name) reappeared to a new life.

Stillman Drake[43] distinguishes among the scholars of the XVI century who engaged in mechanics several traditions: four going back to the Greek classics and two to the medieval authors. At the last, he divides the Italian authors in two groups geographically separate. In the North: Niccolò Tartaglia (1500–1557), Girolamo Cardano (1501–1576), Giovan Battista Benedetti (1530–1590); in the Centre: Federico Commandino (1509–1575), Bernardino Baldi (1533–1617), Guidobaldo Del Monte (1545–1607).

Limiting the range of our interests to those who have dealt with the motion of bodies in an appreciable way, we shall restrict ourselves to Tartaglia and Benedetti. We shall speak further of Guidobaldo Del Monte, when we shall deal with the curve of projectiles.

---

[43]See **Mechanics in Sixteenth—Century Italy**—Selection from Tartaglia, Benedetti, Guido Ubaldo & Galileo—Translated & Annotated by Stillman Drake & I.E. Drabkin—The University of Wisconsin Press, 1969—pp. 5–16.

2.6.1 Niccolò Tartaglia (1500–1557)—His Life and His Works

Niccolò Tartaglia was born in Brescia about the year 1500, became fatherless when six years old and, being of a poor family, had difficulty in continuing the school. According to what he himself says,[44] he remained in half alphabet. When he was twelve years old he was wounded in the face by a soldier of the French army which had occupied the town slaughtering the citizens.

Niccolò had been given for dead, but instead he survived thanks to his mother's care; however he remained maimed, because of the wound to lips and jaw, and having a great difficulty in speaking.

The cognomen Tartaglia (it seems that the true name were Fontana) derived from this speech impediment. The young Niccolò studied alone with determined zeal and, autodidact, became an expert of mathematics and even translated some classical works from Latin, among which also the **Elements** of Euclid. His passion for the mathematics led him to be interested in Algebra and in the most important problem at that time: the solution of the equations of third degree.

We cannot speak at length of his activity of algebraist (that, on the other hand, is what gave him an international fame) and of the mathematical challenges with Cardano and others, since this would bring us out of our field of investigation.[45] We are interested, instead, in those works where he deals with the motion of projectiles and the fall of heavy bodies.

In this regard, we remember that in the XVI century, thanks to some progress made in the fabrication of guns, it was beginning the request of precise rules for the shots of artillery by the artillerymen themselves. The military technology needed the help of mathematicians. Tartaglia devoted himself to this study and, as we shall see later, with regard to the motion of projectiles he was the first to obtain certain results, still correct nowadays. The works which we shall deal with are two: The **Nova Scientia** and the **Quesiti et Inventioni diverse**.

2.6.1.1 The Nova Scientia and the Quesiti et Inventioni Diverse

The first work was published in 1537,[46] but had several editions and reprints (1550, 1581, 1583). Conceived in five chapters (books), as the author stated in the preface, consisted in only three in all the subsequent editions. «Divided into five

---

[44]Most of the biographical notes about him come from autobiographical hints scattered in his works, particularly from **Quesiti et Inventioni diverse** (see the bibliographic reference further on).

[45]The reader can refer to the book (in Italian): Fabio Toscano—**La formula segreta. Tartaglia, Cardano e il duello matematico che infiammò l'Italia del Rinascimento**—Sironi, 2009. A French translation (Belin, Belin edition) is also available.

[46]**Nova Scientia** inventa da Nicolo Tartalea—in Vinegia, per Stephano da Sabio, ad instantia di Nicolo Tartalea brisciano, MDXXXVII.

books: In the First is demonstrated theoretically the nature and effects of uniformly heavy bodies in the two contrary motions that may occur in them, and their contrary effects. In the Second is geometrically proved and demonstrated the similarity and proportionality of their trajectories in the various ways that they can be ejected or thrown forcibly through the air, and likewise the [proportionality] of their distances….»

Then, the third deals with the determination of the distances through the observations and the calculation. On the whole one has to do with a work all devoted to the trajectories of the cannonballs, (nowadays we would say a treatise of artillery), which seems to have originated by the questions put to Tartaglia by an artilleryman of Verona.

Tartaglia dedicates his work to the Duke of Urbino, who later on will ask him for professional advice.

Of course, studying the trajectories of a cannonball means studying the motion of a heavy body tossed with a certain force; since after a certain time the projectile will fall at ground due to its gravity, one must face the problem of considering the comprehence of two kinds of motion, the violent motion and the natural motion, according to the theory of Aristotle. And it is just within the Aristotelian scheme, completely ignoring the impetus-theory, that Tartaglia moves on in his treatment.

Let us look at its fundamental elements. On the acceleration in the fall of heavy bodies:

«FIRST PROPOSITION

Every uniformly heavy body in natural motion will go more swiftly the more it shall depart from its beginning or the more it shall approach its end.[47]»

A uniformly heavy body is

«FIRST DEFINITION

A body is called uniformly heavy which, according to the weight of the material and its shape, is apt not to suffer noticeable resistance from the air in any motion.»,[48]

that is a heavy body according to the everyday language. From the enunciated proportion, one immediately infers that the fundamental element is the distance covered and, then, from this one can deduce that at any instant the velocity of the moveable is proportional to the distance covered.

Notwithstanding the frontispiece of the **Nova Scientia** is constituted by an allegorical picture which represents the ensemble of the sciences as a blockhouse whose entrance is defended by Euclid (that is, in the blockhouse of the science one can enter only if he knows the mathematics), Tartaglia, when describing the

---

[47]**Nova Scientia**, f. 4r. (The translation is taken from **Mechanics in Sixteenth—Century Italy**, op. cit. p. 74).

[48]Ibidem. f. 1r. (**Mechanics …** op. cit. p. 70).

accelerated motion of a heavy body in free fall does not appeal as much to the mathematical rigor as to a simile in the literary tradition:

«This same is also verified in anything that goes toward a desired place, for the more closely it approaches the said place, the more happily it goes, and the more it forces its pace, as appears in a pilgrim that comes from a distant place; for when he nears his country, he naturally hastens his pace as much as he can, and the more so, the more distant the land from which he comes. Therefore the heavy body does the same thing in going toward its proper home, which is the center of the earth; and when it comes from farther from that center, it will go so much the more swiftly approaching it.»[49]

In the subsequent edition of 1550,[50] at this point Tartaglia adds:

«The opinion of many is that if there were a tunnel that penetrated diametrically through the whole earth, and through this there were let go a uniformly heavy body, as said above, then that body when it arrived at the center of the world would immediately stop there. But I say that this opinion (that it would stop immediately upon arriving there) is not true; instead, by the great speed which would be found in it there, it would be forced to pass by with very violent motion, running much beyond the said center toward the sky of our subterranean hemisphere; and thereafter it would return by natural motion toward the same center, and arriving there it would pass by once more, for the same reasons, with violent motion toward us; and yet again it would return by natural motion toward the same center, and pass it still again with violent motion, thereafter returning by natural motion, and so it would continue for a time, passing with violent motion and returning by natural motion, continually diminishing in speed, and then finally it would stop at the said center. By which it is a manifest thing that violent motion is caused by natural motion, and not the reverse; that is, natural motion is never caused by violent motion, but is rather its own cause.»[51]

Therefore, Tartaglia states that the body in free fall in the tunnel which crosses diametrically the Earth is subject to a damped oscillatory motion which at the end will be extinguished in the centre of the Earth itself.

An analogous statement had been already expressed two centuries before, both by Oresme and by Albert of Saxony. One could suppose a direct reading of the work of the later by Tartaglia since, at that time, a Venetian edition of the commentary of Albert to the **De Caelo** of Aristotle already existed.[52]

Stillman Drake substantially espoused this thesis, by attributing to Tartaglia the adoption of the impetus-theory in the version of Albert,[53] that is, with a continuous decrease of the impetus in the course of motion.

Tartaglia, however, limits himself to statements which copy those of Albert, without ever speaking of impetus. This problem will be reconsidered by Galileo in the **Dialogue**, but with the conclusion that the motion lasts up to infinity. We shall have the opportunity of coming back on the subject.

---

[49]Ibidem, f. 4r. (Ibidem, p. 75).

[50]**La Nova Scientia**, Stampata in Venetia per Nicolo de Bascarini a instantia de l'Autore. 1550. See the anastatic reprint by Arnaldo Forni Editore, 1984.

[51]See **La Nova Scientia**, op. cit. f. 4r,v. (**Mechanics ...** op. cit. pp 75–76).

[52]See: Albert od Saxony—**Questiones subtilissime in libros de celo et mundo Aristotelis**—Venice 1492, (ff. 32r–33v)—The quotation is taken from Clagett, op. cit. chap. 9.

[53]See: **Mechanics in Sisteenth—Century Italy**, op. cit., p. 76.

In the second proposition, Tartaglia adds:

«All similar and equal uniformly heavy bodies leave from the beginning of their natural movements with equal speed, but, coming to the end of their movements, that which shall have passed through a longer space will go more swiftly.»[54]

That is, the dependence of the final velocity on the covered distance is reaffirmed. Benedetti, later on, will point out that it is not necessary that the bodies of which we study the fall have the same weight.

On the contrary of what happens for the natural motion:

«A uniformly heavy body in violent motion will go more weakly and slowly the more it departs from its beginning or approaches its end. (Proposition III).»[55]

Moreover, he specifies:

«All similar and equal uniformly heavy bodies, coming to the end of their violent motions, will go with equal speed; but, from the beginning of such movements, that which shall have to pass through the longer space will leave more swiftly.» (Proposition IV).»[56]

And, by insisting on the fact that the two kinds of motion must be considered separately (according to Aristotle),

«No uniformly heavy body can go through any interval of time or of space with mixed natural and violent motion.» (Proposition V).[57]

He further explains, with an example, the impossibility of the coexistence of the two kinds of motion in any point of the trajectory with the fact that the violent motion is decelerated (see the third proposition) while the natural motion is accelerated (see the second proposition).

As a consequence, since a motion cannot be in the same time accelerated and decelerated, it comes that the trajectory of a projectile must be composed by rectilinear tracts and curved tracts. Let us see as he solves the problem:

«Every violent trajectory or motion of uniformly heavy bodies outside the perpendicular of the horizon will always be partly straight and partly curved, and the curved part will form part of the circumference of a circle. » (Supposition II of the second book).[58]

But he is forced to admit that this is an approximate solution. In fact, he continues by saying about a motion of a heavy body which occurs out of the vertical:

«Truly no violent trajectory or motion of a uniformly heavy body outside the perpendicular of the horizon can have any part that is perfectly straight, because of the weight residing in that body, which continually acts on it and draws it toward the center of the world. Nevertheless, we shall suppose that part which is insensibly curved to be straight, and that which is evidently curved we shall suppose to be part of the circumference of a circle, as they do not sensibly differ.»[59]

---

[54]**La Nova Scientia**, op. cit. f. 5r. (**Mechanics ...** op. cit. pp 75–76).

[55]Ibidem, f. 5v. (**Mechanics ...** op. cit. p. 78).

[56]Ibidem, f. 6v. (**Mechanics ...** op. cit. p. 79).

[57]Ibidem, f. 7r. (**Mechanics ...** op. cit. p. 80).

[58]Ibidem, f. 10v. (**Mechanics ...** op. cit. p. 84).

[59]Ibidem, f. 11r. (**Mechanics ...** op. cit. pp 84–85).

Then, contrary to what asserted by some authors in the past, Tartaglia was quite conscious that the trajectory of a projectile consists of a continuous curve of which the representation with rectilinear segments and arcs of circle is only an approximation.

On the other hand, he does not give an explanation of the use of the circle as a curve for approximating the "curved tracts". As it is known, the (uniform) circular motion had always been considered the motion of celestial bodies, even if considered exclusively from a geometric point of view in order "to save the phenomena" by composing a suitable number of circular motions.

In this case, instead, an arc of circle directly shapes a tract of the real trajectory. As we shall see, it will take Guidobaldo and Galileo to realize that the whole trajectory of a projectile consists of a parabola.

An important further result which already appears in the **Nova Scientia** is that, if one changes the elevation of the cannon, the maximum gun-range is obtained for an elevation of 45°. In fact he says in the eighth proposition of the second book:

«If the same motive power shall eject or shoot similar and equal uniformly heavy bodies in different ways violently through air, that shot which shall have a trajectory elevated 45° above the horizon will make its effect farther from its beginning on the plane of the horizon than one elevated in any other way.»[60]

This result is already proudly enunciated by Tartaglia in the dedication of the Duke of Urbino, together with the fact that is possible to reach a target with two different elevations:

«Further, since by evident reasons I knew that a cannon could strike in the same place with two different elevations or aimings, I found the way of bringing about this event, a thing unheard of and not thought by any other, ancient or modern.»

On the motion of the heavy bodies and the trajectory of the projectiles Tartaglia will come back in the subsequent work **Quesiti et inventioni diverse**,[61] even dedicated to Henry VIII (to the Merciful and Invincible Henry VIII, by Grace of God King of England, of France, and of Ireland, etc.).

The work consisted of nine books, but only the first (30 questions on the artillery shots), the seventh (7 questions on the principles of the **Mechanica** of Aristotle) and the eighth (42 questions on the theory of the heavy bodies) concern the subject we are interested with. The first book contains three questions from the Duke of Urbino (dated 1538, i.e. a year after the publication of the **Nova Scientia**). From the point of view of the theory of the motion of bodies, the book does not present new elements, while there are many improvements concerning the technique of the artillery.

---

[60]Ibidem, ff. 16v, 17r. (**Mechanics ...** op. cit. p. 91).

[61]**Quesiti et inventioni diverse** de Nicolo Tartalea Brisciano in Venetia per Venturino Ruffinelli, ad instantia et requisitione, et a proprie spese da Nicolo Tartalea, autore, 1546. The first edition of 1546 was followed by others until the definitive edition **Quesiti et inventioni diverse de Nicolo Tartaglia**—in Venetia per Nicolo de Bascarini, ad instantia et requisitione, et a proprie spese de Nicolo Tartaglia Autore. Nell'anno di nostra salute. MDLIIII. An anastatic reprint of this edition is available with introduction and notes by Arnaldo Masotti—La nuova cartografica, Brescia 1959.

2.6.2 The Mechanics of Giovan Battista Benedetti

The work of Giovan Battista Benedetti, nowadays considered the most important of the immediate forerunners of Galileo, essentially happened in the second half of the XVI century, and has been pointed out to the future historians for the first time by the mathematician and historian of science Guglielmo Libri (1802–1869) in his work **Histoire des sciences mathématiques en Italie**.[62]

Libri devoted several pages of the third tome of his work to commenting the mathematical results obtained by Benedetti and, with regard to the theory of motion, expresses the following judgment:

« ... On l'aurait admiré davantage si l'on avait compris, à cette époque, toute l'importance de sa théorie de la chute des graves, dont on n'a jamais parlé, et qui mérite cependant une place distingué dans l'histoire des sciences.»[63]

and

« ... Benedetti, dont le nome est à la peine prononcé aujourd'hui en Italie, doit être placé au premier rang des savans du seizieme siécle.»[64]

He will be followed by Caverni who starts in the first tome of his **History** with the judgment:

«The science of the motion, made impossible by the errors of Aristotle, remained, one can say, stationary in the books of the ancient Archimedes. Our Benedetti was one of the most valuable in promoting it, by confuting with reasonable arguments those Aristotelian errors, many of which had been shared even by Niccolò Tartaglia himself, for both the natural and the violent motions.»[65]

Caverni was even more eulogistic in the fourth tome where, comparing Benedetti with his forerunners, says:

«But in Giovan Battista Benedetti, from whom a new science begins, the words have a very different sound.»[66] and «The doors of the truth, remained bolted by the peripatetic advices for so many centuries, once so happily made free should lead Benedetti to deliver by his own fair hand to Galileo himself the key for going into the most hidden vestibules of the temple.»[67]

---

[62]Guillaume Libri: **Histoire des sciences mathématiques en Italie, depuis la renaissance des lettres jusqu'à la fin du XVIIe siècle** (Paris 1838–41).

[63]Libri, op. cit. tome troisième, p. 123.

[64]Ibidem, p. 131.

[65]Caverni, op. cit., tomo I, p. 103.

[66]Caverni, op.cit. tomo IV, p. 97.

[67]Ibidem.

Finally, we quote from Vailati, who at the end of the XIX century devoted to Benedetti an important essay:

«Among those who most efficiently contributed to preparing and making possible that great scientific revolution which is marked by the discoveries of the fundamental laws of motion, Giovanni Benedetti ... occupies a special place. The role he had in the first elaboration of the theory and the concepts, which are at the basis of the modern Dynamics, represents a contribution of an entirely different nature than that which, at the constitution of the new science, was brought by the other immediate forerunners of Galileo.»[68]

2.6.2.1 The Life

The biographical data on Giovan Battista Benedetti (Venice 1530–Turin 1590) are exceedingly scanty: some of them inserted by himself in a work of him,[69] some others supplied by his contemporary Luca Gaurico (1475–1558) in his **Tractatus astrologicus ...** (Venice, 1552).

According to that is told, one knows that he did not have preceptors and after he was seven years old he did not attend any school and, as he himself remembers, only Niccolò Tartaglia taught him the first four books of Euclid.

«... Besides, since one must give back to everyone what is his own, for it is right and legitimate, Niccolò Tartaglia read to me only the first four books of Euclid, all the rest was investigated by my private study and work: in fact nothing is difficult to be learned by a willing man.»[70]

Benedetti remained in Venice until about 1558 and then passed to Parma called there by Duke Ottavio Farnese as a professor (*Lettore*) of philosophy and mathematics. He remained at Parma about eight years also dealing with astronomic studies. He took leave from Parma at the beginning of 1567 and passed to Turin invited there by the Duke of Savoy Emanuele Filiberto. Here, besides holding the office of the mathematician of the Court, was also teacher at the University; he remained in Turin, even after the death of Emanuele Filiberto, until his own death in 1590.

---

[68]Giovanni Vailati: **Le speculazioni di Giovanni Benedetti sul moto dei gravi**—Atti della R. Accademia delle Scienze di Torino, vol. XXXIII, 1898 (Reprinted in: Giovanni Vailati—**Scritti** a cura di Mario Quaranta—Arnaldo Forni Editore, 1987—vol. II, pp 143–160).

[69]**Resolutio Omnium Euclidis Problematum Aliorumque ad hoc necessario inventorum una tantummodo circini data apertura, Per Joannem Baptistam De Benedictis inventa**. Venetiis MDLIII.

[70]Taken from the eighth page (the pages are not numbered) of the dedication to abbot Gabriele Guzman of the **Resolutio**: «...ceterum quia cuique quod suum est reddi debet, nam pium et iustum est, Nicolaus Tartaleas mihi quattuor primos libros Euclidis solos legit, reliqua omnia privato labore et studio investigavi: volenti namque scire nihil est difficile».

2.6.2.2 The Works and the Successes of the Ideas on the Mechanics

Exhausted these scanty biographical data,[71] we think it is important, and particularly meaningful, to account for (besides the works) also how his ideas were received by his immediate contemporaries, in order to be able to weight later on the impact that the work of Benedetti has had on those who succeeded to him in the scientific milieu.

Benedetti has dealt with many subjects, although sometimes only occasionally, but we are only interested in what regards the mechanics.[72]

In the long (21 pages) dedication of his first work,[73] to abbot Gabriel Guzman, Benedetti anticipates his criticisms to Aristotle, with regard to the laws of fall of heavy bodies, which he shall develop in the two subsequent editions of the work **Demonstratio Proportionum Motuum Localium contra Aristotelem et Omnes Philosophos** (Venetiis MDLIIII).

The anticipation of his ideas on the motion of heavy bodies has been attributed to the fear to be expropriated of his ideas by other people. Indeed, this happened with the content of the **Demonstratio**. In fact, while there is no evidence of the diffusion of the work among the contemporaries in the immediacy, eight years after a plagiarism of a certain Taisner[74] was published.

It was a work which contained inside, and announced in the subtitle, no less than the **Demonstratio proportionum motuum localium, contra Aristotelem & alios Philosophos.**

In this way Benedetti's ideas went into the scientific debate through that plagiarism which had a spread greater than the original. In fact, Stevin himself (by quoting their false author) reported them in his work **Elementa hydrostatica** (1586).

---

[71]Still today, the most complete study on this regard is given by the memoir of Giovanni Bordiga: **Giovanni Battista Benedetti filosofo e matematico veneziano del secolo XVI**— Atti dell'Istituto Veneto di Scienze, Lettere ed Arti, Tomo LXXXV, Parte seconda, pp 585–754 (1925–1926)—reprinted in 1985, with a bibliographic updating by Pasquale Ventrice on the occasion of the workshop on «Giovan Battista Benedetti e il suo tempo». In addition one can consult the entry of him (by Stillman Drake) in the vol. 1 (1970), pp. 604–609, of the **Dictionary of Scientific Biography** op. cit.

[72]With regard to this, the most important studies are due to Carlo Maccagni (of whom we quote **Le speculazioni giovanili "de motu" di Giovan Battista Benedetti** (Pisa, Domus Galilaeana, 1967) and to Enrico Giusti (of whom we quote **Gli scritti «de motu» di Giovan Battista Benedetti**—Bollettino di Storia delle Scienze Matematiche, vol. XVII (1997) fasc. 1, pp 51–103.

[73]See footnote 69.

[74]For this, see: C. Maccagni **Le speculazioni giovanili «de motu»** ... op cit. pp XXVII–XXXII. The book of Maccagni also contains excerpts of the dedicatory letter of the **Resolutio omnium Euclidis problematum** and the text of the two editions of the subsequent work **Demonstratio** ....

Let us begin to see which are the ideas on the motion that Benedetti introduced in the dedication to abbot Guzman attributing to him the invitation to do it:

«... Once, when we were still together, you eagerly begged and besought me to write something on the subject of natural motions based on sound theory and also, as far as possible, supported with mathematical demonstrations.»[75]

Benedetti absolutely wants to have assured to him the priority of the ideas where one demonstrates that the theory of Aristotle which maintains that the velocity of a falling heavy body is proportional to the weight of the body itself and in inverse proportion with the density of the medium is groundless.

Benedetti, in his exposition, refers to both the fifth book of Euclid's **Elements** and the work **On Floating Bodies** of Archimedes. He begins, as already Tartaglia had done in the **Quesiti et Inventioni diverse**, by establishing, for the uniform and homogeneous bodies, the proportionality between weight and volume. The relevant demonstration is done for a body which is three times another: almost always the demonstrations are done on particular examples and not in general. Then he moves on to deal with the fundamental subject that is the fall of heavy bodies. Substantially, he maintains the proportionality between the falling velocity and the density (or better, the specific weight), having deducted the buoyancy.[76] Using modern symbols, given two bodies A and B, of densities d_A e d_B, which fall in a medium of density d, and called V_A e V_B their velocities, one has[77]

$$\frac{V_A}{V_B} = \frac{d_A - d}{d_B - d}.$$

It must be remembered that Benedetti, as obviously all the people of that time, cannot think to measure the velocity in the sense that we attribute to these words nowadays. The essential element was the fall time: the body which covered the same distance in a smaller time, or took the same time to cover a greater distance, was the swifter.

Giusti says with regard to the **Resolutio**: «In it, essential conceptual novelties occur: a new law for the motion of heavy bodies which, contrary to the generally accepted Aristotelian law, accounts for the fact that a body with its specific weight equal to that of the medium does not move neither downward nor upward and leads to the isochronism of the fall of the homogeneous bodies.»[78]

As we have already mentioned, the **Demonstratio** had two editions at close range one from the other, starting from Carlo Maccagni—to whom their bibliological study is due—denoted as *Demonstratio and **Demonstratio.

---

[75]*Olim cum adhuc una essemus, magno me opere orasti obsecratusque es aliqua de motibus naturalibus speculatione sollicita conscriberem, idem quantum possibile est Mathematicis demonstrationibus muniens.»*

[76]The demonstration, based on the Archimedes' principle had been perhaps suggested by the reading of the first treatise of Archimedes translated by Tartaglia in 1551.

[77]E. Giusti, op. cit. in 72) p. 60.

[78]Ibidem, p. 69.

In the *Demonstratio, which is an arrangement of the ideas anticipated in the Resolutio, it is particularly pointed out that in the void the isochronism in the fall is not limited to homogeneous bodies, but is valid for all heavy bodies, independently of their density. In the **Demonstratio, which, according to Giusti, has at last a less general formulation than *Demonstratio, the densities are no more alluded and only the weights appear (deducted of the buoyancy). If A and B are two equal bodies of different species, which fall into two media m_1 and m_2, one has

$$\frac{V_A}{V_B} = \frac{P_A - P_1}{P_B - P_2},$$

where P_1 e P_2 are the weights of an equal volume of the two media.[79]

In the **Demonstratio, the fact that the resistance of the medium is not only due to buoyancy, but depends on the surface of the body as well, is also reported. This indicated that the equality of the velocities of fall for homogeneous bodies of the same material should hold only in the void. Obviously, for dealing with this fact, Benedetti has not at his disposal the indispensable mathematical technique, therefore he should limit himself to deal only with particular cases.

After the Resolutio and the two Demonstratio, Benedetti went back to be concerned with the motion of heavy bodies in his last work published in Turin in 1585: Diversarum Speculationum Mathematicarum & Physicarum Liber— Taurini MDLXXXV.

The work is organized in six parts of which the third (De Mechanicis), the fourth (Disputationes de quibusdam placitis Aristotelis) and, in part, the sixth (Physica Mathematica responsa per Epistolas) regard the mechanics.[80]

In this work, Benedetti proposes that the velocity of fall of the heavy bodies were proportional to their weight in the medium and in inverse proportion to the intrinsic resistance, that is, their surface.

As Giusti remarks,

«The path of Benedetti then ends with a position substantially Aristotelian. The proposition he has many times rejected, according to which the velocity of fall was proportional to the weight and in inverse proportion with the resistance of the medium, constitutes the last achievement of his researches. The only difference is: where Aristotle spoke of the absolute weight, Benedetti considered the weight in the medium; and whereas according to the philosopher of Stagira the resistance of the medium was proportional to its density, in the opinion of the Venetian mathematicians it does not depend on the greater or smaller density of the medium, but only on the surface of the body which moves onto it. The differences are lesser than what the vehemence of the polemic could suggest».[81]

---

[79]Ibidem, p. 75.

[80]Excerpts of this work and also of the Resolutio, and the text of **Demonstratio are translated in English in the book Mechanics in Sixteenth—Century Italy by Stillman Drake & I.E. Drabkin—The University of Wisconsin Press, 1969.

[81]E. Giusti, op. cit. p. 94.

Always remaining on the subject of the motion of heavy bodies, we must add in conclusion an important remark of Benedetti, obviously still in opposition to Aristotle. In chap. XXIV of the **Disputationes de quibusdam placitis Aristotelis**, with regard to the acceleration of the motion of fall of the bodies Benedetti says:

«Now Aristotle should not have said *(De Caelo* I, Ch. 8) that the nearer a body approaches its terminal goal the swifter it is, but rather that the farther distant it is from its starting point the swifter it is. For the impression is always greater, the more the body moves in natural motion. Thus the body continually receives new impetus since it contains within itself the cause of motion, which is the tendency to go toward its own proper place, outside of which it remains only by force.».[82]

Therefore, the acceleration of falling bodies is sustained by increases of the impetus subsequently supplied ad infinitum. But Benedetti did not give a mathematical formulation of this.

Caverni, Wohlwill and Vailati said words of great appreciation for what we have recalled above, to the extent that Wohlwill described Benedetti as "der bedeutendste"[83] among the immediate forerunners of Galileo.

There are not evidences that he was aware of the medieval developments of the kinematics, but this has not prevented Duhem from considering him as an epigone of the Parisian school.[84] This opinion was shared by Koyré as well,[85] although the text quoted from him induces to think more to the use of an expression generally accepted (the impetus) than to the conscious acceptance of a particular theory:

«First, every heavy body, when moved either naturally or by force, receives on itself an impression and impetus of motion, so that, even if separated from the motive force, it moves by itself for some length of time. (Indeed, if it is set in *natural* motion, it will always increase its velocity: for then the impetus and impression [of motion] are always being increased, since the motive force is always joined to the body.) Thus, if we move the wheel with our hand and then remove the hand from it, the wheel will not immediately come to rest but will turn for some length of time.»[86]

---

[82]**Diversarum Speculationum**, op. cit. p. 184—«Aristoteles 8 cap. primi lib. De coelo, dicere non deberet quanto propius accedit corpus ad terminum ad quem, tanto magis fit velox; sed potius, quanto longius distat a termino a quo tanto velocius existit quia tanto maior sit semper impressio, quanto magis movetur naturaliter corpus, et continuo novum impetum recipit cum in se motus causam contineat, quae est inclinatio ad locum suum eundi, extra quem per vim consistit. » (Translation from **Mechanics ...** op. cit. p. 217).

[83]"The most authoritative".

[84]P. Duhem: **Études sur Léonard De Vinci**—troisième série—op. cit. pp 214–227.

[85]Alexandre Koyré: **À l'Aube de la science classique**, Paris, Hermann—1939, pp 41–54.

[86]«Nempe omne corpus grave, aut sui natura, aut vi motum, in se recipit impressionem et impetum motus, ita ut separatum a virtute movente per aliquod temporis spatium ex seipso moveatur. nam si secundum naturam motu cieatur, suam velocitatem semper aget, cum in eo impetus et impressio semper ageantur, quia coniuctam habet perpetuo virtutem moventem. Unde manu movendo rotam, ab eaque eam removendo rota statim non quiescet, sed per aliquod temporis spatium circumverteretur.» **Diversarum Speculationum ...** op. cit. pp 286–287. (**Mechanics ...** op. cit. p. 230).

This passage is excerpted from the last section (**Physica Mathematica Responsa per Epistolas**) of the **Diversarum Speculationum** where there are some letters of Benedetti answering questions regarding physical subjects asked by important persons of whom he had made the acquaintance.

The passage quoted above belongs to one of the three letters addressed to Giovanni Paolo Capra, a gentleman of Novara; in this case, the question was related to the motion (rotation) of the well-wheel on its axis.

Benedetti explains the causes because of which the motion of rotation of wheel caused by hand is extinguished after a certain time that the hand has left the wheel.

The same problem had been already dealt with in chap. XIV of the section **De Mechanicis** and to that part Duhem refers (see footnote 84), praising Benedetti for having continued and completed the work of Buridan and Albert of Saxony.

Indeed, the whole work of Benedetti gives the impression of having been written more as an opposition to the Aristotelian traditions than as a criticism, or continuation of what maintained by his more or less immediate predecessors. It is quite true that, in that time, it was not usual to quote the authors when using or discussing their results, except when one had to do with a direct and contingent polemic (in this way also Galileo will behave), the only explicit references being to the ancient authors.

In the case under examination, considering the unceasing and exclusive reference to the "philosopher" it seems more correct to exclude a reference to Buridan. Also the fact that Buridan himself had previously dealt with the problem of the motion of the wheel (see quotation in footnote 39) seems more simply due to the circumstance that quite that problem was one of those which more usually occurred to those which were studying the motion of bodies.

For a demonstration that, if not the history, at least a certain propensity of the historians recurs, we must note that, in the case of Benedetti, to the obsession of Duhem of wanting him to be a follower of the Parisian school, the doggedness of his successors succeeded of attributing to Galileo a direct dependence on the work of Benedetti.

The action begins, we can say, with the **History** of Caverni, who, by forcing the (attested) historic truth, heavily fictionalizes some little facts the testimonies of which have been left over.

Let us make a digression to introducing the subject.

In 1589 (at the express wish of Grand Duke Ferdinando Dei Medici) Jacopo Mazzoni of Cesena, one of the most famous humanists of that time in Italy, renowned above all for his studies on Dante, was called as a professor at the University of Pisa. Mazzoni, in Pisa, gave ordinary lectures of Aristotelian philosophy and extraordinary lectures of Platonic philosophy (we are using the locutions of that time).

In the same year, also the twenty-four years old Galileo was arrived at Pisa, as a professor of mathematics. As far as it is known, the young Galileo entered the intellectual community that had been formed around the forty-years-old Mazzoni, a successful and prestigious intellectual and, as we would say nowadays, an interdisciplinary scholar. Of the association and the friendship with Mazzoni, Galileo talked both to his father and to Guidobaldo Del Monte.[87]

The stay of Galileo at Pisa lasted only three years; in fact, he moved to Padua in September 1592. In 1597, Mazzoni published the first volume (destined to remain the sole) of a work where the philosophies of Plato and Aristotle were compared.[88]

Galileo, as received the book in Padua, wrote a long letter to the old colleague in which, besides the compliments for the book he promises to continue to read, he shows «a greatest satisfaction and comfort» in seeing Mazzoni «in some of those questions that in the first years of our friendship we discussed together with great pleasure, (now) to incline to that part believed true by me, and the contrary by You ....».

Those discussions were on the Copernican system and Galileo takes the opportunity for devoting almost the whole letter to that subject. In fact he ends by saying «... I do not want to trouble you anymore, but only to ask you to tell me, if you agree, if in this matter it is possible to save Copernicus. I'm tired of writing and you of reading: so by removing all the slownesses of ceremonies I shall end by kissing your hands ...».[89]

At this point, we must say that Mazzoni in his work quotes in different parts the **Diversarum Speculationum** of Benedetti considering it as a reference text for the physical questions.

It is this circumstance that has sparked off all considerations on a possible indoctrination of Galileo by Mazzoni by using the book of Benedetti as a textbook. All this should have taken place in the triennial 1589–1592, in the Pisan period of Galileo.

Let us see how Caverni reconstructs the facts: «There was, among the young auditors in Pisa in those times, also Galileo, and since Mazzoni recognized to him a peculiar attitude of the mind to penetrate the science of motion, recommended to him the book of Benedetti and privately explained its speculations. The young disciple, from those words of the Master and from the reading he suggested, felt instilling in him the first ineffable taste of freedom in thinking, and since the fervent advices and efficacious example had driven him to no more trust in the Aristotle's

---

[87]From Galileo to his father Vincenzio (… I'm keeping very well and attend to study and to learn with Signor Mazzoni, who greets you. And not having anything also to tell, I end. From Pisa, the ninth of October 1590…). From Guidobaldo Del Monte to Galileo (… I rejoice at the fact that you are getting along well with Signor Mazzoni, not without envy from me, who would sometimes be with both and take pleasure of his talks: give my best regards and a long hand-kissing to Signor Mazzoni …. From Monte Barroccio, the eighth of December 1590 …) E. N. X, pp 44–45.

[88]Jacopo Mazzoni: **In Universam Platonis, et Aristotelis Philosophiam Praeludia, sive de comparatione Platonis & Aristotelis**—Venetiis, MCXCVII. There is a critical edition of this book edited by Sara Matteoli and with an introduction of Anna De Pace—M. D'Auria Editore—2010.

[89]E. N. II, pp 193–202.

authority, then, he concluded, not even in that of other philosopher, not excepted Benedetti himself, if he too should be considered to deviate from the rectitude of the natural truths.»[90]

We shall reopen the question later on, when we shall deal with the writing of Galileo of the Pisan period left handwritten and later called with the name **De Motu antiquiora**. In any case, it must be remarked that Caverni, in his free historical reconstruction, takes care to give notice that in the Pisan period Galileo's ideas were not a slavish repetition of those of Benedetti.

2.7 Galileo and the Engineers of the Renaissance

The title of this short section follows that of a book famous for having been one of the first, if not the first, to deal with the development of the several techniques in the Renaissance, trying to give of them an overall outline.[91]

However, we do not really want to deal with the world of the "engineers". The subject concerns us here only for what deals with a possible influence that the variegate world of the "mechanical arts" has exerted on Galileo and his work. Therefore we anticipate here some considerations which would have equally find place in the course of the subsequent chapters. But we prefer to join them to the discourse done with regard to Benedetti and, above all, of Tartaglia. As we know, Galileo, after his installation at the University of Padua, was obliged to deal also with subjects which had nothing to do with the "pure science", but rather with the military architecture and connected problems.[92]

Between the XV and the XVI century in Italy (but, in part, also in the rest of Europe) an environment had been developed constituted by artists, designers of various sorts, inventors, who can be grouped (following Gille) under the generic label of "engineers". It is clear that the first name which springs to mind is that of Leonard, but we could add Francesco di Giorgio Martini, Leon Battista Alberti, preceded by Lorenzo Ghiberti, Filippo Brunelleschi etc. In Germany, we can mention Albrecht Dürer.

In the period we generally denote as Renaissance, besides a rebirth of the "humanae litterae" also an exceptional development of the technique had happened, at the beginning also favored by the translations of the ancient works (on the machines and their use) passed from the Greek to the Arabic world. Later on, new treatises, written in Italian, were printed, full of pictures representing new machines. The authors of these treatises did not restrict themselves to describe machines really constructed, but more often, were projects not actually carried out, but elaborated in the pictures in minute details. It's enough to browse, for instance, a copy of the **Diverse et artificiose macchine** of Agostino Ramelli (1588).

---

[90]R. Caverni: **La Storia del Metodo Sperimentale in Italia,** op. cit. Tomo IV, p. 275.

[91]Bernard Gille: **Les ingénieurs de la Renaissance** (Hermann, Paris, 1964).

[92]See, in E. N. II, the **Breve instruzione all'architettura militare** and the **Trattato di fortificazione**.

The protagonists of this development of the technique did know neither the Greek nor the Latin, i.e. were, as at that time were named, men without letters.

The problem of the use of the "vernacular Italian" (or Tuscan) had received a great interest also by the men of letters (see the **Dialogo delle lingue** of Sperone Speroni—1549) and therefore the Italian was legitimate anywhere, except for the academic teaching, where the Latin remained the official language.

Nonetheless, the first work that Galileo personally saw into print in the Paduan period (**Le Operazioni del Compasso geometrico et militare**—1606),[93] as "lecturer of mathematics in the Study of Padua", as he presented himself in the title page, was written in Italian.

It is evident that he had taken into account the final-users to whom the work was addressed. On the other hand, the high regard in which he held the world of the technicians clearly appears at the beginning of his last work, the **Discourses**:

> «*Salviati.* Frequent experience of your famous arsenal, my Venetian friends, seems to me to open a large field to speculative minds for philosophizing, and particularly in that area which is called mechanics, inasmuch as every sort of instrument and machine is continually put in operation there. And among its great number of artisans there must be some who, through observations handed down by their predecessors as well as those which they attentively and continually make for themselves, are truly expert and whose reasoning is of the finest.
>
> *Sagredo.* You are quite right. And since I am by nature curious, I frequent the place for my own diversion and to watch the activity of those whom we call "key men" *[Proti]* by reason of a certain pre-eminence that they have over the rest of the workmen. Talking with them has helped me many times in the investigation of the reason for effects that are not only remarkable, but also abstruse, and almost unthinkable.»[94]

The first of the quoted passages is the beginning of the dialogue in the role of Salviati, alter ego of Galileo, whereas the second is the intervention into the dialogue by the Venetian gentleman Giovanfrancesco Sagredo, but Favaro says «... we are sure that Galileo wanted to refer to himself what that adds to the said things ...».[95]

All this for anticipating the observation that Galileo shall ever choose the language in which to express himself taking into account both of the addressee and the diffusion he shall want to have. He was conscious of moving in a world which was not anymore the world of Benedetti.

Galileo could talk also to the "engineers".

---

[93]Besides in E. N. II, see **Il Compasso geometrico e militare di Galileo Galilei** edited by Roberto Vergara Caffarelli, Edizioni ETS, 1992 and Galileo Galilei—**Operations of the Geometric and military compass**—Smithsonian Institute Press, 1978. (Facsimile reprint translated with an introduction by S. Drake).

[94]E. N. VIII (**Discorsi e Dimostrazioni matematiche intorno a due nuove scienze**), p. 49. (Drake p. 11).

[95]Antonio Favaro: Galileo **Galilei e lo studio di Padova, II**, reprint of the original work of 1883, Editrice Antenore—Padova, 1996, p. 70.

Bibliography

We quote here only books and papers not mentioned in the footnotes of the chapter.

J. Burckhardt, *The civilization of the renaissance in Italy* (Phaidon Press, London, 1945)

S. Drake, Impetus Theory and quanta of speed before and after Galileo. Physis **16**, 47–65 (1974)

E. Gilson, *La Philosophie au Moyen Age, des Origines patristiques à la Fin du XIV Siècle* (Paris, Payot, 1944)

É. Gilson, *Medieval essays—cascade books* (2011) (first French edition, Vrin-Paris, 1986)

J. Le Goff, *Intellectuals in the middle-ages* (Wiley Blackwell, New York, 1993) (first French edition, 1957)

E. Grant, Motion in the void and the principle of inertia in the middle ages. Isis **55**, 265–292 (1964)

E. Grant, Bradwardine and Galileo; Equality of Velocities in the Void. Arch. Hist. Exact Sci. **2**, 344–364 (1965)

E. Grant, *The Foundations of Modern Science in the Middle Ages* (Cambridge University Press, Cambridge, 1996)

C.H. Haskins, *Studies in the History of Medieval Science* (Cambridge Harvard University Press, Cambridge, 1927a)

C.H. Haskins, *The Renaissance of the 12th Century* (Cambridge Harvard University Press, Cambridge, 1927b)

C.H. Haskins, *Studies in the History of Medieval Culture* (Oxford at the Clarendon Press, Oxford, 1929)

J. Huizinga, *The Autumn of the Middle Ages* (Translated by R. J. Payton and U. Mammitsch). (University of Chicago Press, 1997) (first Dutch edition, 1919)

W. R. Knorr, *Ancient Sources of the Medieval Tradition of Mechanics.* (Istituto e Museo di Storia della Scienza, Firenze—Monografia N. 6, 1976)

R. Marcolongo, *La Meccanica di Leonardo da Vinci* (Napoli S. I. E. M., 1932) (Memoir extracted from vol. XIX, Serie 2ª, N. 2 of Atti della R. Accademia delle Scienze fisiche e matematiche di Napoli)

E.A. Moody, M. Clagett (eds.), *The Medieval Science of Weights* (The University of Wisconsin Press, Wisconsin, 1952)

E. A. Moody, *Studies in Medieval Philosophy, Science and Logic. Collected Papers 1933–1969* (University of California Press, California, 1975)

R. Pasnan (ed.), *The Cambridge History of Medieval Philosophy* (Cambridge University Press, Cambridge, 2010)

F. Purnell Jr., Jacopo Mazzoni and Galileo. Physis **14**, 273–294 (1972)

P. Riché, *Les Écoles et l'Enseignement dans l'Occident chretien de la Fin du Ve Siècle au Milieu du XIe Siècle* (Aubier Montaigne, 1979

C. Truesdell, *Essays in the History of Mechanics* (Springer, Berlin, 1968)

Part II
Galileo and the Motion

For greatest convenience of the reader, we insert here a concise chronology relative to the principal occurrences in the scientific life of Galileo. Among the numberless biographies of Galileo, the reader may look up the following books: Stillman Drake: **Galileo at Work. His scientific Biography**—University of Chicago Press, 1978; John L. Heilbron—**Galileo**—Oxford University Press, 2010. We remind here only the dates deemed useful as a support to the text of the book.

1564	Born in Pisa on February 15
1581	He enters the University of Pisa at the Faculty of Arts to study Medicine
1585	He goes back to Florence, where the family had returned since 1574, without having taken a degree. He continues the studies of mathematics, already privately undertaken following his own inclination, during the last university years
1587	First visit to Rome. He meets with Christopher Clavius at the Collegio Romano
1588	He begins to correspond with Christopher Clavius and Guidobaldo Del Monte
1589	On July, he is appointed to the chair of mathematics at the University of Pisa, with the help of Guidobaldo del Monte
1589–92	He writes the Dialogue and the Treatise (**De Motu**)
1592	He moves to the University of Padua as professor of mathematics
1592–1609	He runs academic courses at the University of Padua. The first course (short version) on Le **Mecaniche** belongs to the academic year 1592–1593. The second one (long version) has been held in the academic year 1598–1599. He gives lectures on the fortifications, on the **Euclid's Elements**, on the Ptolemy's **Almagest**, on the sphere, on the Planetary Theories, etc.
1597	He constructs the geometric and military compass. It is the year of the famous letter to Jacopo Mazzoni. Also, he confirms to Kepler his agreement to Copernican model
1602	Letter to Guidobaldo del Monte on the oscillations of the pendulum
1604	Correspondence with Paolo Sarpi on the law of fall of the heavy bodies
1606	He publishes **Il compasso geometrico e militare**

1609	Letter to Antonio De' Medici on the parabolic trajectory of the projectiles. Construction of the telescope
1610	Discovery of the Jupiter's satellites and consequent publication of **The starry Messenger (Sidereus Nuncius)**. He becomes Chief Mathematician and Philosopher of the Grand Duke Cosimo II of Tuscany
1611	Second visit to Rome. He is elected to the Accademia dei Lincei
1612	He publishes the **Discourse on the Bodies that stay atop Water or move in it (Discorso intorno alle Cose che stanno in su l'Acqua o che in quella si muovono)**
1613	He publishes the **Letters on the Sunspots (Istoria e Dimostrazioni intorno alle Macchie Solari)**
1623	He publishes **The Essayer (Il Saggiatore)** dedicated to the Cardinal Maffeo Barberini, just elected Pope with the name of Urbano VIII
1625	He begins to work at the book which will become the **Dialogue concerning the two chief World Systems (Dialogo sopra i due massimi sistemi del mondo)**
1629	After a long interruption, he takes again to work to the **Dialogue**
1630	He ends the writing of the **Dialogue**
1632	The **Dialogue** is printed. By October, he is ordered to stand trial for heresy
1633	Trial of Galileo in Rome. On June 22nd, he is pronounced "vehemently suspected of heresy" and condemned to formal imprisonment. He will live at Siena (as a guest of the Archbishop Ascanio Piccolomini) before returning to his Arcetri home under house arrest. Also, the permission of leaving Arcetri for taking care of his health will be refused to him
1635	The writing of the **Discourses (Discorsi e dimostrazioni matematiche intorno a due nuove scienze)** begins
1636	Negotiation with Louis Elzevir for the publication of the work in Holland
1637	He works at the fourth Day of the **Discourses**
1638	On April, he works at the dialogue on the force of the percussion. On July, the Discourses (four Days) are published at Leiden. On December, he dictates what will be called "sixth Day" (on the force of percussion)
1639	Vincenzio Viviani establishes himself at Arcetri in Galileo's home. He takes dictation (Galileo is now completely blind) for the considerations on the theory of proportions of Euclid (it will be, in future, the fifth Day of the Discourses) and collaborates on various demonstrations
1641	Also, Evangelista Torricelli arrives at Arcetri and writes from dictation some considerations on the Euclid's theory of proportions
1642	On January 9th, Galileo dies

Chapter 3
The Young Galileo and the de Motu

In the spring of 1585 Galileo left the University of Pisa without having taken a degree (in medicine), and a long period passed (until November 1589), before he went back there as professor of mathematics, in which he devoted himself to the study and to the private teaching. A writing of a dialogue (in Latin) between two persons, one of whom (Alexander) is the spokesman of Galileo and the other (Dominicus) is the man who puts the questions, belongs to this period, or to the beginning of the following, and has as subject problems regarding the motion.

The dialogue was left unfinished and the subject was resumed in a treatise on the same argument in the initial period of the Pisan teaching (an unanimous agreement among the scholars regarding the date of writing of both the dialogue and the treatise has not been reached yet).

Both writings were never published during Galileo's lifetime, even if their drafting in Latin makes one think that the author had originally planned them for a possible publication.

We shall see, later on, the probable reasons which led Galileo to keep them as manuscripts.

3.1 On the Editions of the Dialogue and the Treatise

Let us now deal with the combination of circumstances which led to the publication of the dialogue and of the treatise, since we think that there is something to learn even from this.

The publication of tomo XI of the **Opere complete di Galileo Galilei**, edited by Eugenio Albèri[1] dates back to 1854. The volume we are talking about

---

[1] **Le Opere di Galileo Galilei**—Prima Edizione completa condotta sugli autentici Manoscritti Palatini e dedicata a S. A. I. e R. Leopoldo II Granduca di Toscana—Tomo XI—Firenze, Società Editrice Fiorentina, 1854.

© Springer International Publishing Switzerland 2016
D. Boccaletti, *Galileo and the Equations of Motion*,
DOI 10.1007/978-3-319-20134-4_3

contained (from pp. 9 to 80) the **Sermones de Motu Gravium** (published for the first time), whose title was, obviously, due to Albèri. Under this title, which referred to the content, the editor had placed a dialogue (pp. 9–55), followed (pp. 56–90) by four chapters belonging, as we shall see, to a treatise on the laws of motion, and by a fifth chapter already used by Galileo in his last published work, **Discorsi e dimostrazioni Matematiche** ….

Let us omit, for now, dealing with the "Avvertimento" put before the **Sermones** by Albèri.

In the last quarter of the century after the Unity of Italy (we are still referring to the XIX century), one began to think about a National Edition of Galileo's work. Finally things were carried out, above all by means of Antonio Favaro who was also the editor of the whole work together with Isidoro Del Lungo as literary assistant.

In 1890 the first volume appeared.[2] Besides writings which was named **Juvenilia** (substantially, notes of the young Galileo on philosophical arguments) and writings regarding the study of the centres of gravity and the hydrostatics (and others clearly due to the study of Archimedes' works), the volume contained a section (pp. 257–419) entitled **De Motu**, preceded by five pages of "Avvertimento" by the editor Antonio Favaro.

Since that moment (1890), but actually since some years before (when Favaro had anticipated the publication of the manuscripts found in the National Library of Florence), a pure "querelle" begins regarding the date of composition and therefore the chronological sequence of writings which compose the section entitled **De Motu**.

According to the order in which Favaro had arranged the manuscript, the section began with a treatise ("De Motu") of twenty three chapters (in Favaro's opinion most probably unfinished) followed by the remaking of the first two chapters (by I.E. Drabkin classified as "second version"[3]), in turn followed by the remaking of more ten chapters—including the first two ("third version" in Drabkin's opinion). Finally the "Dialogue", already published by Albèri, and some notes (later classified "Memoranda"), concluded the section.

The "querelle" mentioned above came to involve the most important Galilean scholars: between '800 and '900, A. Favaro and E. Wohlwill, later on A. Koyré, L. Olschki, I.E. Drabkin, S. Drake, T. B. Settle and, finally, R. Fredette, W. E. Hooper, M. Camerota, E. Giusti (by quoting only a few). The author of this book is in concordance with the chronological arrangement (firstly suggested by

---

[2]**Le Opere di Galileo Galilei—Edizione Nazionale**, Volume I. Firenze, Tipografia di G. Barbera—1890. (From now on, quoted as E. N. I).

[3]See: **Galileo Galilei: on Motion and on Mechanics**—Comprising **De Motu** and **Le Meccaniche** (translation of I.E. Drabkin and Stillman Drake)—The University of Wisconsin Press, Madison, 1960.

Fredette[4]—1969—and then widely demonstrated by Giusti[5] —1998): Dialogue—Treatise—Second version—Third version. It seems to be very reasonable the splitting of the Treatise in two books (Fredette).

At this point, the reader who is not a member of the community of experts may wonder about the why of the "querelle" which, moreover, was quite heated. The problem was, and still is (in the history of science whatever solution is never completely definitive) the fundamental understanding of what the first ideas of Galileo on the law of motion have been and in what way they evolved until the work of his maturity **Discorsi e Dimostrazioni Matematiche** ..., which founded the modern mechanics.

At this point it is time to ask ourselves how and when the manuscripts (labeled Ms. Gal. 71 in the National Library of Florence) arrived in the place where they are conserved at disposal of the scholars.

3.2 The Vicissitudes of the Manuscripts

We begin the story by calling to speak the man who owned the manuscripts of Galileo on his master's death, that is Vincenzio Viviani, who cured Galileo in his last years (starting from October 1639).

In 1674 Viviani says, listing what was in his possession:

«The writings of the aforesaid inventory consist (apart from some speeches and letters of others) of proofs of Galileo's printed works, or in speeches, and letters of him, which are already known (*si vedono fuori sparse*); and among Galileo's things, whose copies, as far as it is known, do not exist, there are only two. The first one is a manuscript of Galileo consisting of several small quinternions in octavo entitled on the cover **De Motu Antiquiora**, which is recognizable as one of his first youngish studies, and from which one sees that since that time he could not force his free intellect into the conventional philosophizing of the ordinary Schools. But what is particularly meaningful, scattered in this manuscript, was properly inserted by him in the right places in the works he published, as one can see. The other ....».[6]

The conclusions of Viviani in the last phrase are not completely exact (as it has been demonstrated), but this is not interesting to us; instead let us follow the fate of the "small quinternions" after the death of Viviani, occurred in 1703.

«Signor Viviani had collected, both from the heirs of Galileo and from others, as many works (*monumenti*) of his master he could find; but in order to remove them from the inquisition of the bigoted people, since he himself was fallen under suspicion of being an irreligious person, he kept them hidden at home in a wheat pit. One day this house was

---

[4]See: R. Fredette: **Galileo's De Motu Antiquiora**—Physis, XIV, 321–348, 1972—and, preceding, PhD thesis (Université de Montréal—1969).

[5]E. Giusti: **Elements for the relative chronology of Galilei's De Motu Antiquiora**—Leo S. Olschki Editore, 1998.

[6]**Quinto Libro degli Elementi d'Euclide** ovvero Scienza Universale delle proporzioni spiegata colla dottrina del Galileo con nuovo ordine distesa, e per la prima volta pubblicata da Vincenzio Viviani ultimo suo Discepolo .... In Firenze, Alla Condotta, MDCLXXIV, pp 104–105. (Our translation).

inherited by abbot Jacopo Panzanini, nephew (ex sorore) of Viviani, and at his death in 1737, the pit was open at intervals and many bundles of the aforesaid writings were moved or sold to shopkeepers for enveloping "quidquid chartis amicitur ineptis".»[7]

Let us see the continuation (with some additional detail), in the narration of Giovanni Targioni Tozzetti, physician, great naturalist, and trustworthy scientist of the Gran Duke:

«After the death of abbot Jacopo, which occurred in 1737, someone, I do not know who, from time to time opened that wheat pit, dug up a bundle of writings of Galileo, and brought them to be sold by weight to a certain Cioci, grocer in the market.

In the spring of 1739, it happened that the distinguished doctor Gio. Lami, by his habit, went to dine in a villa of him, at the Osteria del Ponte delle Mosse, with several friends, and passing through the market, he prompted Sig. Gio. Batista Nelli, later senator and knight, that it had been better to purchase from Cioci (the grocer) a certain quantity of mortadella, that was credited to be better than any other.

In fact they entered the shop, and the senator ordered to have sliced two "lirate" of mortadella, and put the envelop in the hat. Just arrived at the tavern, they asked for a plate for laying the mortadella and in that occasion the Sig. senator realized that the sheet used by Cioci to wrap the mortadella was a letter of Galileo.

He removed the grease from the sheet with a napkin the best he could, then folded it up and put it in his pocket, without telling anything to Lami; in the evening, once gone back to town, and said goodbye to him, he dashed to the shop of Cioci, from whom he heard that at intervals an unknown male servant sold a bundle of that type of writings to him.

Then, he bought back the sheets which were left in Cioci's hands, with the promise that, if other sheets ended up in his hands, he would have kept them and they would have discovered where they came from. In fact, within a few days a greater bundle arrived, and Sig. senator came to know that they came out of the aforementioned wheat pit, therefore in 1750 with a few scudi he managed to have in his hands all the remainder of the precious treasure, which was buried there for years.

On the other hand, many bundles had been already scattered before, in enveloping "quidquid chartis amicitur ineptis" … Sig. senator Nelli, once purchased the manuscripts, has rearranged them and has made greatest studies on them and, as once he was so kind to tell me, has written an extensive and reasoned life of Galileo, and of the most distinguished disciples of him, to be printed in conjunction with many of their posthumous works and the letters; but who knows when his many political activities will allow him to do it?.»[8]

---

[7]**Memorie e Lettere inedite finora o disperse di Galileo Galilei**, ordinate ed illustrate con annotazioni dal Cav. Giambattista Venturi—Parte Seconda—Modena, MDCCCXXI. (Our translation). The quoted excerpt belongs to the Preface. Venturi reconstructs the history of the manuscripts. For the continuation, yet we have preferred to quote directly the source, from which, also comes the Latin phrase («all the wares waste paper's used to fold»). It has to do with the last verse of the first epistle of the second book of Horace's Epistles (we have reported John Conington's classical version). The habit of Latin quotations is as old as the hills.

[8]**Notizie degli Aggrandimenti delle Scienze Fisiche accaduti in Toscana nel corso di anni LX del secolo XVII—Raccolte dal dottor. Gio. Targioni Tozzetti—Firenze MDCCLXXX**. Tomo primo, pp 124–125. (Our translation). The work of Targioni Tozzetti, in three volumes, contains an extraordinary quantity of information on the science in Tuscany in the XVII century (as he promises in the title) and also the celebrated Magalotti's "**Saggi di Naturali Esperienze**" i.e. the report of the experiments of the Accademia del Cimento.

Later on, Clementi Nelli did it (by publishing his work at Lausanne in 1793) and mentioned the possession of the manuscripts, but without talking about the envelope of mortadella:

> «I also hold some studies made by the Florentine Philosopher in his youth, and by him transcribed in several quinternions, in one of which one can find written **De Motu Antiquorum** etc.; in others some errors contained in the works of Aristotle are pointed out. Most of this content is reported in the works of Galileo printed until now».[9]

Supporting the version of Targioni Tozzetti, Venturi intervenes, still in the preface to the second part of his work quoted above:

> «Like this Sig. Targioni narrates; neither Sig. Nelli, who usually contradicts him, disprove the news, and only says of having purchased these manuscripts from a junk dealer and from others, to whom they were transferred by the heirs Panzanini.»

Later on, all the collected manuscripts were assembled by the Grand Duke and subsequently lodged at the National Library of Florence.[10]

The novella is of those at happy ending and the historians of the science must be grateful to the Florentine gentlemen who, in occasion of a hike, have helped to save a bundle of Galileo manuscripts. To safeguard their class of gourmet it must be specified that, at that time, the mortadella was produced artisanally with a long ripening and, at the end, cooked. It was the most tasty product which was obtained from the swine: it was a food for wealthy persons.[11]

3.3   The Dialogue

> The dynamics of the Leaning Tower Experiment, to which Galileo gave an "Archimedean" mathematical formulation in his Pisan dialogue on motion, was not the dynamics of Bradwardine or Buridan, nor of their fourteenth and fifteenth century followers. E.A. Moody –loc. cit. p. 410

The dialogue takes place at Pisa, at Bocca d'Arno, in winter season. Alexander meets Dominicus who is walking quickly and addresses him «where are you going, my dearest Dominicus?» and convinces him to stop and then to go on their way talking.

---

[9]**Vita e Commercio Letterario di Galileo Galilei** scritta da Gio. Batista Clemente De' Nelli—Losanna, 1793—Volume II, p. 759. (Our translation).

[10]A. Procissi (1959–1985): **La Collezione Galileiana nella Biblioteca Nazionale di Firenze**—Roma, Ist. Poligrafico e Zecca dello Stato.

[11]The most ancient recipe of the mortadella can be found in the book of the chef of the Court of the ducal family of Este, Cristoforo di Messisbugo: **Libro novo nel qual s'insegna a' far d'ogni sorta di vivanda** (Venetia, 1557) (reprinted by Forni Editore, 2008).

Dominicus after some preambles, in which he also alludes the lectures of Girolamo Borri[12] (on the heavy and light bodies), maybe followed by Galileo when he was a student of medicine, begins with a series of questions (six).

The questions concern the motion and Dominicus asks Alexander «for I know that on this subject you will either say nothing or bring forth something new and very near the truth itself.».[13]

The questions are:

«First, whether you believe it true that at the turning point of motion a rest is required.

Second, what cause do you allege for this, namely that, if two bodies equal in size, of which one is, for example, of wood, the other of iron, so that one is heavier than the other, are let down at the same time from a certain very high place, the wood is carried through the air more swiftly than the iron, that is, the lighter more swiftly than the heavier if, indeed, you admit this as true.

Third, how it happens that natural motion is faster at the end than at the middle or at the beginning, but violent motion is faster at the beginning than at the middle, and faster there than at the end.

Fourth, why the same body goes down more swiftly in air than in water; and, indeed, why certain bodies go down in air, which are not submerged in water.

Fifth (this is a request of our very dear friend Dionisio Font, the most worthy knight) what cause do you give for the fact that guns, both cannons used against fortifications as well as manual arms, throw lead spheres farther along a straight line if they project them at right angles to the horizon, than if on a line parallel to the same horizon, although the first motion is more opposed to natural motion.

Sixth, why the same guns throw heavier balls more swiftly and farther than lighter ones, as those of iron in comparison with wooden ones, although the lighter ones resist the impelling force less.»[14]

Dominicus asks questions to Alexander:

«Now since you have grown accustomed to very reliable, very clear and also very subtle mathematical demonstrations, as those of the divine Ptolemy and the most divine Archimedes, you cannot in any way give your approval to cruder arguments: and since these things which I have proposed to you are not very far removed from mathematical considerations, it is with eager ears that I expect something beautiful from you.»[15]

As we shall see, during the dialogue Alexander will answer only three of the questions, since the work will be left unfinished (Galileo will start again to deal with the same subjects, in the treatise, some years later).

It is immediately clear that Alexander wants to emphasize his disagreement with Aristotle on some fundamental points. In fact, he states in advance that, for

---

[12]Girolamo Borri (or Borro), professor of philosophy at the University of Pisa until 1585, a few years before had also published his book: Hieronymus Borrius Arretinus, **De Motu Gravium et Levium**, Florentiae, MDLXXVI.

[13]«... scio enim te in hac materia aut nihil dicturum, aut aliquid novi et veritati ipsi propinquissimum in medium allaturum.» E. N. I, p. 368 (29–31).

[14]Ibidem (5–27) (Translation by R. Fredette, as for the other excerpts of the Dialogue).

[15]Ibidem (31–36).

expressing his opinion about the proposed problems, he must emphasize three things: the mover, the moveable and the medium through which the motion occurs. The last two are the same both in the natural motion and in the violent motion: the former, i.e. the mover, is not the same in both motions, in fact in the natural motion it is its heaviness or lightness; in the violent motion (the mover) is a certain impressed power (*virtus quaedam impressa*).

This is the fundamental point on which Galileo (Alexander) founds his criticism to Aristotle: a body which moves with violent motion, released by the hand or by any other agent which has pushed it, does not move because of the medium with the complicate mechanism of the antiperistasis, but because of a force impressed to it with the initial push and which will decrease during the motion until vanishing.

This is the mechanics of Avempace and, even before, of Philoponus.

Obviously, given the division of the tasks in the well tested mechanism of this kind of dialogue, the reluctant Dominicus, a meek predecessor of the more famous Simplicio of the **Dialogue** of the maturity, will be perplexed by the strength of the examples displayed by Alexander. The last of these, in the opinion of the author the most convincing (*potissimum argumentum*), is that of the motion of a sphere rotating around a diametral axis put on two hearings.

Let us quote the original passage:

> «consider a marble or iron sphere, perfectly round and smooth, which can be moved on an axis at rest on two pivots; then let a mover come near who twists both ends of the axis with his finger tips: surely in that case the sphere will rotate for a long time: and yet the air has not been put into motion by the mover, nor can the medium ever come up into the parts abandoned by the mobile, since the sphere never changes place.

> What, then, is to be said of this violent motion? by what will the sphere be moved when it is outside the hand of the mover? what is to be said, except that it is moved by an impressed force?».[16]

But just the "*potissimum argumentum*" makes the opposition of Dominicus begin. In fact, Alexander has touched a dangerous subject: the circular motion. Let us remind what states Aristotle with regard to this in **On the Heavens** (see I, 8, 276a-277b). For him the circular motion is reserved to celestial spheres, is endless and does not require any force at all for being maintained.

In fact Dominicus intervenes:

> «I cannot not surrender to your arguments: and yet concerning this last one, which you seem to deem outstanding, there is something which I could doubt. For those who defend the contrary viewpoints could perhaps reply to an argument of this kind by saying that that motion is not violent, since it is circular. For since natural motion is contrary to violent motion, but to circular motion there is no contrary, circular motion will in no way be violent: and since it is not violent, the consequence that you deduce from the motion of the sphere, will be of no moment.»[17]

---

[16]E. N. I, p. 372 (15–23).

[17]Ibidem (24–31).

Alexander replies by distinguishing between the vertical violent motion, opposite to the natural motion, and the motion of a projectile thrown obliquely with respect to the horizon line. In this case we shall have to do with a mixed motion, that is consisting of violent motion and natural motion.

The motion of the sphere will be of this type: by rotating, some parts of the sphere go down whereas other parts go up and, moreover, since the sphere with the axis weighs on the hinges, we shall also have a resistance.

Finally, the motion of a sphere like this could be non-violent only in the ideal case in which the sphere would be at the centre of the world and would rotate just around to the centre of the world itself. But on this supposed "circular natural motion" an uncertainty remains: being non-violent, this motion should last forever, but perhaps this does not fit the nature of the Earth.

At this point, Dominicus declares that he is convinced that the violent motion is not due to the medium (with the antiperistasis) which, on the contrary, prevents it, but to an impressed force; however he wants an explanation about the way the moveable is moved by the heaviness or by the lightness in the natural motion.

Therefore, once concluded the discussion on the violent motion and on its mover, the impressed force, one passes to the natural motion and its mover, the weight.

Whereas in the first case Alexander goes back to Avempace (even if he does not mention him), in the second he goes back to Archimedes, who instead is quoted. With plenty of examples and arguments, Alexander, i.e. Galileo, fresh from studies on the works of Archimedes, explains the concept of specific weight and that which nowadays we just call Archimedes principle.

The heavy and the light with their related hierarchies do not exist:

«I say, then, that one does not speak of the heavy and the light except in a comparison.»[18]

As we shall see, this concept will be further reaffirmed in the treatise. The fundamental criticism to Aristotle (which is obviously due to having taken up the physics of Archimedes) is that the light bodies do not exist for themselves but only as bodies differently heavy. As regards the motion upward and downward of a body in a medium, he finally concludes:

«From this it is evident in general that those bodies which are heavier than the medium through which they must be carried are carried downward; but those which are lighter than the medium through which they must be carried are carried upward.»[19]

Of course, Alexander, obliging, illustrates with explanatory examples all questions put by Dominicus for the various cases where, we would say, one applies the Archimedes principle.

---

[18]«Dico ita, gravia et levia non dici nisi in comparatione» E. N. I, p. 378 (3).

[19]«Ex quo universaliter patet, corpora illa deorsum ferri, quae medio, per quod ferri debent, graviora sunt; ea vero sursum ferri, quae medio, per quod ferri debent, leviora fuerint» E. N. I, p. 384 (34–36).

At last, Dominicus, satisfied asks:

«But what about my problems?»[20]

Only one of your problems, answers Alexander, has clearly solution according to what we have exposed about the media, the moveables and the motion; the explanation of the others will come soon, for both the already established considerations and those we shall do.

The problem is that of the fourth question put at the beginning by Dominicus, that is, why a moveable gets down more swiftly in the air than in the water. The answer is evident, Alexander says: since the ratio between the gravity of the body and that of the air is greater than the ratio between the gravity of the moveable and that of the water, the consequence is that the moveable gets down with greater force (*maiori vi*)[21] in the air than in the water.

Galileo (Alexander) agrees, without objecting, to what the question considers true and granted (the difference in the velocities of fall of a body—always the same—in media of different densities), but, with respect to Aristotle, he has already introduced a new element: it is not the weight (that absolute, as if the body was weighted in the void) to determine the velocity of fall in a medium like air, water etc., but the relative specific weight with regard to the medium.

And, up to here, it is Archimedes who inspires him but for the rest, unfortunately, it is still Aristotle who lays down the law.

The question of Dominicus concerned the "velocity" of the fall, in fact it was used the comparative adverb "*citius*". The answer of Alexander says that a body falls with greater force (*maiori vi*) in the medium for which the ratio is greater between the gravity of the body and the gravity of the medium itself.

Then, from this, two consequences come for Alexander; the first that the velocity is proportional to the force (weight), the second that the velocity of fall in a medium is constant (being constant the weight of the body).

With regard to the first, one can then conclude that bodies with different specific weight fall (in the same medium) with different velocities.

With regard to the second, which foresaw a uniform motion, there was no experimental proof that could disprove this conclusion. Almost certainly, since true and reliable observations were lacking, all was based on the everyday experience of the fall of heavy bodies, in the air, from relatively short distances.

Got an answer to the fourth question, Dominicus presses: «And the question on the turning point of the motion?».

For answering the question, Alexander begins by demonstrating that, in the violent motion, the force impressed by the mover weakens successively and there are never two points where it can have the same value (*duo puncta, in quibus eiusdem roboris sit virtus impellens*). From this, one can deduce that, in the moveable thrown

---

[20]E.N.I, p. 389 (8–9).

[21]E. N. I, p. 389 (17–18).

upright, since the impressed force will always weaken during the motion whereas the gravity will remain constant, there will be an instant in which the impressed force will pass through the value of the gravity. But it will not stop at this value, as neither at other values too, therefore it does not exist an instant of rest as Aristotle maintained, saying that before the moveable fell with natural motion an interval should occur because the motion could not be violent and natural in the same time.

Dominicus appears satisfied with the explanation but, since Alexander in the confutation of Aristotle had alluded the void, asks him to tell something.

Alexander answers that on the void he could tell many things, but he will omit them in order not to go too far away from the prearranged purpose; he will only tell the things which depend on what he has already exposed. I shall limit myself, he says, to demonstrate that in the void the motion, on the contrary of what Aristotle asserts, cannot be instantaneous.

Since a heavy body falls in a medium when its heaviness exceeds that of the medium itself, in the case of void (which has vanishing heaviness) the excess will be given by the whole heaviness of the body, which is finite, and therefore the motion of the body will not be instantaneous, but with a finite velocity.

Again Dominicus appears satisfied, but asks for further explanations about the errors of Aristotle, particularly as what regards the resistance that the medium opposes to the fall of a heavy body.

In this case, Alexander replaces the ratio (of Aristotle) between power and resistance ("geometric proportion") with the difference ("arithmetic proportion"). This solution, as we have already seen, corresponds to the theory of Avempace.

He finally ends the argument by demonstrating that the velocity of fall of different bodies, of the same matter but of different weights, is the same. In fact, we have already seen that the magnitude which determines the velocity of fall is the specific weight and not the absolute weight of a body.[22]

At this point, Dominicus, satisfied with the explanations from Alexander as usual, asks for going back again to the solution of the leftover problems, that he waits for pricking up his ears.

Now, answers Alexander, you will receive the solution of the problem, among those which can be deduced from the only things we have said before.

The problem is that in which the cause is investigated of the fact that the natural motion is swifter at the end than at half, and here is swifter than at the beginning

---

[22]We remind, with regard to this, the account of Viviani (nowadays called into question by the majority) were it was told that Galileo maintained «... that the velocities of the moveables of the same matter, but with different weights, on moving in the same medium, do not keep otherwise the proportion of their weights, now assigned to them by Aristotle, but the (moveables) move all instead with the same violence, proving what repeated experiments performed from the summit of the campanile of Pisa...». (Our translation). To the best of our knowledge, an English translation of the "Historical Account of the life of Galileo Galilei" does not exist. Besides E. N., XIX, one can refer to the following recent editions: Vincenzio Viviani—**Vita di Galileo** a cura di Bruno Basile—Salerno Editrice, 2001, p. 38 and **Vita di Galileo** di Vincenzio Viviani (con appendice di testi e documenti) a cura di Luciana Borsetto—Moretti & Vitali editori—Bergamo, 1992, p. 87.

(i.e. the third question). The problem is therefore to understand why the motion of fall of heavy bodies is accelerated.

The explanation of Galileo is rather tortuous and, judged a posteriori, partly contradictory. We have seen until here that, according to Galileo, the motion of fall of a heavy body is uniform with a velocity proportional to its specific weight. Why, instead, Dominicus talks of it as an accelerated motion?

Galileo says that, if we throw a body upward, we must do it with an impressed force greater than its weight. During the ascent, the impressed force will be decreasing until being equal to the weight; from this point onwards the body will come down and the impressed force (now smaller than the weight) will be always decreasing and then the velocity of the body always increasing, because the force opposing to the weight will gradually diminish until vanishing.

From this point onwards, according to Galileo's theory, the body should proceed with uniform motion.

But, Dominicus asks, what does it happen if the body starts from rest, instead of having been thrown upward before?

Alexander answers that a body which starts from rest, before being left to fall, was anyway held up by a hand or by a support and then by a force which opposed (equal and contrary) the weight. Because of that the two cases are equal.

Alexander asserts that, even if we let a stone fall from a high tower (*Si enim ab alta turri lapis descendat*), the motion remains, however, accelerated during all the path because the height of the tower is not enough to drain the impressed force completely. In addition, there is also an effect, so to say, of perspective. To a man who is looking standing up in front of the tower, the body will seem to cover equal distances in shorter and shorter times since it is seen under decreasing angles.

In any case, Alexander had premised in advance that, even if what he was asserting happened contrary to the opinion of many, for him this should not be important provided it would be consistent with the reason and the experience. But he adds that this time the experience shows the contrary.

The dialogue ends with a further demand of specifications from Dominicus who asks if things change in case the heavy body is not only left to fall but even pushed by strength downward. The answer is that the action of this force, opposite to that impressed, speeds up its vanishing and, then, determines a swifter passage from the accelerated motion of fall to the uniform one.

Three of the questions put by Dominicus at the beginning remain unanswered. At the present time, the scholars who agree with the idea that the Dialogue is prior to the Treatise, from this deduce the obvious consequence that Galileo interrupted it because of not being satisfied with the result obtained.

As we shall see, in the Treatise, he will also deal with the subjects held over in the Dialogue but the results will be disappointing the same. With the two ingredients, the "vis impressa" and the Archimedean hydrostatics, he will not yet succeed to construct a satisfactory dynamics.

3.4 The Treatise

At the end of those rambling notes ("appunti scuciti"), as Favaro calls them, at p. 418 and 419 of the first volume of the E. N., it appears what Fredette interprets as a programme of a work on the motion.[23]

This conjecture is more than reasonable. And it seems also reasonable, starting from a phrase of Chap. 14, to infer that the Treatise had been conceived by Galileo as consisting of two books, the first of them comprising the first thirteen chapters. Subsequently, he would have rewritten the first two chapters (the second version according to Drabkin) and, finally, the first ten (always including the first two), which is the third version according to Drabkin.

The list of the subjects, in the substance, reflects what will be effectively discussed in the Treatise (and partly has been discussed in the Dialogue), but we think it is interesting all the same to quote it integrally in order to control in case, a posteriori, how much of the programme has been effectively carried out and also the sequence of the subjects, which is not a secondary detail.

Here is the list, with the warning that the gaps are due (as Favaro attests) to the zones where the paper of the manuscript was worn out:

«It may be asked whether heavy bodies really move toward the center. On this, Ptolemy, *Almagest I*, Chap. 7.

Will the impressed force be consumed by time or by the weight of the body?

By what is natural motion caused?

By what is forced motion caused?

Is a medium necessary for motion?

Is there an absolutely heavy and an absolutely light?

Are the elements in their proper place heavy or light?

On the ratio of the [speeds of] motions of the same body in different media.

On the ratio of the [speeds of] motions of different bodies in the same medium.

On the cause of the slowness and speed of motion.

Is there [an interval of] rest at the turning point?

Is natural motion always accelerated and why is it accelerated?

Is slowness and speed of natural motion due to rareness or <density of the medium>?

In motion three items are considered: the moving body, the medium, and the motive force.

Of what help or hindrance is the shape of moving bodies to their motion?

The ratio of the weights of the same heavy body in different media, on which depends the question of the ratio [of the speeds] of their motions.

---

[23]See: R. Fredette: **Galileo's De Motu Antiquiora—Notes for a reappraisal**—Eurosymposium Galileo 2001, p. 174.

If the weight of the medium and the speed of the body are known, the weight of the body is also known. If the weight of the body and of the medium are known, the speed of the motion is also known.

If the speed and weight of the body are known, the weight of the medium is also known.

On circular motion.

To be considered is the ratio [of the speeds] of motions on inclined planes; and whether it happens that lighter bodies fall more swiftly at the beginning, just as, in the case of the balance, the smaller the weights, the more easily the motion takes place.

The medium retards natural motion in this way: for example, when a bell falls, it is, so to speak, a solid body consisting of air enclosed by the metal, and so it is lighter than if air were not present.»[24]

The last two phrases, which we omit, are difficult to be reconstructed, because of the too many gaps.

A part of this programme, as we have seen, has been dealt with and discussed in the Dialogue.

Obviously, at the basis of all, there are the problems which have their roots in the questions put by the ancient Greek philosophers and of which, in his opinion, Aristotle gave the solution.

The fundamental questions concerned: the relation between the velocity of fall of the heavy bodies, their weight and the density of the medium; the cause of the non-natural motions and their kinematical features.

In all the three versions, the Treatise begins with two chapters acting as an introduction, where general elements are exposed: for comparing the gravity of bodies of different matter, one must refer to equal volumes; the Providence has distributed the matter in the universe in the most rational way by concentrating the heaviest matter at the centre. At the centre there is also the Earth.

This is in agreement with the Ptolemaic conception, which is quoted in the first subject of the programme, reminded with a marginal note in the first version of the Treatise, and direcly quoted in the two subsequent versions.

By the way, we take the opportunity for a couple of reflections, even if not directly pertaining to the themes of the Treatise. First: from the repeated mentions and quotations, relevant to Ptolemy, in both the Dialogue and the three versions of the Treatise, it can be reasonably inferred that, until the time of the third version, Galileo was convinced of the Ptolemaic conception.

Second: from the text from the **De Revolutionibus** he quotes in Chap. 7 of the first version, concerning the fact that it is possible to obtain a rectilinear motion by composing two circular motions, one can infer that he had read the work of Copernicus thoroughly.

In fact, one has to do with an argument hidden enough (Lib. III, cap. III— *Quomodo motus reciprocus sive librationis ex circularibus constet*). Therefore, when he was writing the Treatise, he had read carefully Copernicus, but was not yet Copernican. During the three–year period of the Pisan teaching, he was instead

---

[24]E. N. I, pp 418–419. (Galileo Galilei **On Motion and on Mechanics**, op. cit. pp 130–131).

already become Copernican, as the letter to Jacopo Mazzoni of 1597 (E. N., II, pp. 193–202)—where he reminds the discussions of that period—testifies.

Coming back to our subject, we say that it is possible to group the subsequent chapters of the Treatise into two sections: one regarding the natural motions (Galileo continues to call them as Aristotle did), i.e. the motions due to the weight of the bodies, and the other regarding the forced or violent motions, i.e. due to an external force.

As to the former, in the second chapter there is already an important element of novelty (with respect to Aristotle): the elimination of the concept of "lightness". One of the points listed in the programme was «*is there an absolutely light?*» and, in the Dialogue, it was insisted in the assertion that one cannot speak of heavy and light except for comparison. The second chapter of the Treatise in the first version has the title «*The heavy substances are by nature located in a lower place, and light substances in a higher place and why*».

Finally, Chap. VII, third version, has the title «*That no upward motion is natural*» followed by Chap. VIII entitled «*That contrary to Aristotle's view, no body is without weight*» (*Gravitatis corpus nullum expers esse, contra Aristotelis opinionem*).

The detaching from the Aristotelian doctrine on this point is definite and drastic, strengthened by examples and demonstrations. Bodies light for themselves do not exist, only bodies more or less or equally heavy than others do exist.

3.4.1 The Natural Motions

The natural motion is therefore, exclusively, that of the bodies downward (i.e. towards the centre of the Earth) and is due to their intrinsic gravity. In the eighth chapter Galileo deals with the problem of the motion of different bodies in the same medium, by examining the various cases (equality and/or difference of kinds and sizes in their possible combinations), for demonstrating what have been the errors of Aristotle in **On the Heavens**.

Let us take the more evident case. In book IV of **On the Heavens**, Aristotles asserts that «*magnum aurum citius ferri quam paucum*»,[25] i.e. a bigger piece of gold goes down swifter then a smaller one.

Clearly, there is nothing more ridiculous, affirms Galileo by contesting the "philosopher". In the following, he exposes three examples to show the absurdity of the Aristotelian text; one of them is the case of two stones (one twice the other) left to fall from a high tower.

Can it happen that when the smaller is at half of the path, the other has already reached the ground? Galileo presses. But, he adds, for making always use of reasoning, instead of examples, since we are looking for the causes of effects not

---

[25]E. N. I, p. 263 (13). «that a large piece of gold moves more swiftly than a small piece».

reported by the experience, we shall expose our thought from whose confirmation one will have the failure of Aristotle's opinion. Therefore, we say that moveables of the same kind, even though they may differ in size, are moved with the same velocity and a bigger stone does not fall swifter than a smaller one.[26]

Then Galileo continues by proposing to join two bodies of the same kind (one bigger and the other smaller). If what Aristotle says was true, the whole of the bodies should fall in a medium with an intermediate velocity (smaller than that of the bigger body and greater than that of the smaller body), since the body which falls with the smaller velocity slows down that which falls with greater velocity and that which falls with greater velocity speeds up that which falls with smaller velocity. But this contradicts the fact that the whole of the two bodies (of the same kind) is heavier of the bigger body and therefore, according to Aristotle, should fall with a velocity greater than that of the bigger body when it falls alone.

The same reasonings, reversed, can also be applied to the case in which the bodies (of the same kind) go up, instead of going down (having a specific weight smaller than that of the medium).

Then he states what are the ratios between the velocities of bodies of different matter but of the same volume which fall in the same medium.

The ratios between velocities will be equal to the ratios between the quantities of which the gravities of the bodies exceed with respect to the gravity of the medium. From the above, it is not difficult to calculate the proportions between the velocities of moveables of different kinds which fall in different media.

Of course, in the case of the bodies which go up, one proceeds in the same way, but now what is proportional to the velocity of the body is the excess of the specific weight of the medium with respect to that of the body.

Then Galileo concludes by saying:

«These, then, are the general rules governing the ratio of the speeds of [natural] motion of bodies made of the same or of different material, in the same medium or in different media, and moving upward or downward.»[27]

Except to be obliged, immediately, to admit that the experience does not confirm the rules he has enunciated. It is not here the place of investigating the causes of these divergences (in any case they happen "per accidens"),[28] he says. As a matter of fact, always referring to the example of a heavy body which falls from a high tower, he admits that the problem is always that of understanding why the motion of fall of a heavy body is slower at the beginning. The accelerated motion of the fall of heavy bodies, as we know, will be "his" problem for years to come.

We can, until here, summarize the "theory" of Galileo concerning the "natural motion" (in a medium) in the following way:

---

[26]Ibidem (25–31).

[27]E. N. I, p. 273 (19–21).

[28]Ibidem (28–29).

1. The fundamental element relevant to the natural motion (i.e. due to gravity) of a body is not its weight but its specific weight.
2. In the case of a motion in a medium, one will have to compare the two specific weights (of the body and the medium: if the specific weight of the body is greater than that of the medium, the body will fall downward; if it happens the inverse, the body once immersed will be pushed upward.
3. The velocities, of fall or of ascent, will be proportional to the difference of the two specific weights. Since, in any case, this difference is obviously a constant, the consequence is that the motion (of fall or ascent) will be a uniform motion.

As for the theory of Aristotle it is usual to represent the velocity of fall through the formula $V = K\frac{P}{M}$ (where P is the weight of the body and M the resistance of the medium), in the case of Galileo instead one can represent it as $V = K(P - M)$ (where now P is the specific weight of the body and M that of the medium).

Of course, it is a purely symbolic way of representing the things, since we have not at our disposal, for the magnitudes about which we are speaking, the necessary definitions. In the language of Galileo, in the first case one has a geometric proportion and in the second an arithmetic proportion.

We finally remark that Galileo does not say anything regarding the point listed in the programme: «*Of what help or hindrance is the shape of moving bodies to their motion?*». On this, as we have seen, Benedetti stranded as well. The mathematical tools at their disposal were obviously unequal to the task.

In the subsequent chapters, until to the thirteenth included, Galileo is engaged to contest what stated by Aristotle about the void. As we have seen some time ago, Aristotle denied the existence of the void and, above all, maintained that in it the motion would have been impossible. One of the reasons was that, being the resistance of the medium vanishing (since one was doing with the void), the motion would have been possible only at an infinite velocity (and then instantaneous), and this was absurd.

Galileo, on the basis of his "arithmetic proportion", and then with the velocity of the moveable directly proportional exclusively to its gravity, can instead assert that, since the gravity remains finite, also the velocity will be finite and then the motion possible.

The curious thing, in this question, is that the Galileo at that time, inter alia, blamed Aristotle for what he himself would have sustained some years later. In fact, he says:

«Now, in the first place, Aristotle errs in that he does not prove that it is absurd for different bodies to move in a void with the same speed.»[29]

Once established that the motion in the void is possible, Galileo ends:

«In which, in opposition to Aristotle and Themistius, it is proved that only in a void can differences of weights and motions be exactly determined.»[30]

---

[29]«Et, primo, Aristoteles peccat, in hoc, quod non ostendit quomodo absurdum sit, in vacuo diversa mobilia eadem celeritate moveri ...» E. N. I, p. 283 (20–22).

[30]E. N. i, p. 294 (15–16).

Also in this question, Galileo appears in numerical examples of his "arithmetical proportions". At the beginning of the second book (according to Fredette) or, anyhow, in the fourteenth chapter of the Treatise, he at last deals with a "non-Aristotelian" problem: the motion of a body along an inclined plane.

One has still to do with a natural motion, but which does not drive the body directly toward the centre of the Earth following the vertical. Further, always relating the velocity of fall to the weight of the body, Galileo also resolves to find what is the ratio (*proportio*) between the motions of the same body along planes of different inclinations (and, when he says ratio between the motions, he means ratio between the velocities).

The problem of the inclined plane was already dealt with in the ancient times by both Hero (in the Treatise **Mechanics**, where the "simple machines" were described) and Pappus (about III century A. D.) in the eighth book of his **Collectiones Mathematicae**.[31]

The work of Hero remained yet unknown until the end of XIX century and therefore, obviously, could not have exerted its influence on the medieval and Renaissance mechanics.

On the contrary, the work of Pappus was known, and, even better, a Latin translation by Federico Commandino had only just been published in 1588. But the solution given by Pappus of the problem of the motion along an inclined plane did not appear completely convincing to the scholars of the Renaissance. In short, one can say that, according to Pappus, if one indicates with C the force necessary to displace the body of weight A (usually, a sphere) on a horizontal plane, and with α the angle that the inclined plane forms with the horizontal plane, force F, necessary to move the sphere on the inclined plane, results to be

$$F = C + C\frac{\sin\alpha}{1 - \sin\alpha} = \frac{C}{(1 - \sin\alpha)},$$

where the modern symbols have been used.[32]

Set in these terms, from the formula immediately comes out the inadequacy of the solution given by Pappus. In fact, with the increasing of angle α (until 90°), force F tends to infinite.[33]

Besides having not correctly broken up the force acting on the body, Pappus has also taken for granted that for displacing the body on the plane it is necessary to exert a force proportional to the weight of the body itself. Error which will not be repeated by the "protoinertial" Galileo.

Before him, the problem was also dealt with by Gerolamo Cardano in his work **Opus Novum de Proportionibus numerorum** ... (Basilea, 1570). But Cardano

---

[31]For both works, for the part concerning the inclined plane, see: **A Source Book in Greek Science** by Morris R. Cohen and I.E. Drabkin—Harvard University Press, 1966, pp 194–200.

[32]Ibidem, p. 196. See also: Galileo Galilei—**Le Mecaniche** –Edizione critica e saggio introduttivo di Romano Gatto, Leo S. Holschki Editore, 2004, pp XXIV–XXVIII.

[33]Contrarily to the correct formula $F = A \sin\alpha$, which, for $\alpha = 90°$, gives the weight of the body, i.e. the force which pushes the body along the vertical.

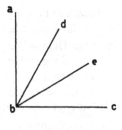

Fig. 3.1

mistook as well, considering that it was necessary to take into account a force for displacing the body on the plane: a force for overcoming the resistance of the medium.[34]

Anyhow Galileo, with a preamble which reminds what will be, more than forty years after, the incipit of the third day of the **Discourses**, asserts being the first to deal with the subject:

«The problem we are now going to discuss has not been taken up by any philosophers, so far as I know. Yet, since it has to do with motion, it seems to be a necessary subject for examination by those who claim to give a treatment of motion that is not incomplete. And it is a problem no less necessary than neat and elegant.

The problem is why the same heavy body, moving downward in natural motion over various planes inclined to the plane of the horizon, moves more readily and swiftly on those planes that make angles nearer right angle with the horizon; and, in addition, the problem calls for the ratio [of the speeds] of the motions that take place at the various inclinations.

The solution of this problem, when first I had tried to investigate it, seemed to require explanations that were by no means simple. But while I was examining it more carefully, and was trying to analyze its solution into its basic principles, I finally discovered that the solution of this problem, as of others which at first glance seem very difficult, depended on known and obvious principles of nature.

We shall begin now by setting forth these ideas, since they are needed for our explanation of the problem. And first, so that everything may be better understood, let us explain the problem by an example.

Let there be a line *ab* (see Fig. 3.1) directed toward the center of the universe and thus perpendicular to a plane parallel to the horizon. And let line *bc* lie in that plane parallel to the horizon. Now from point *b* let any number of lines be drawn making acute angles with line *bc*, e.g., lines *bd* and *be*.

The problem, then, is why a body moving down descends most quickly on line *ab;*and on line *bd* more quickly than on *be,* but more slowly than on *ba;* and on *be* more slowly than on *bd.* We must find, furthermore, how much faster the body descends on *ba* than on *bd,* and how much faster on *bd* than on *be.* Now to be able to answer these questions, we must first take into consideration what we noted above, namely (as is quite clear), that a heavy body tends to move downward with as much force as is necessary to lift it up; i.e., it tends to move downward with the same force with which it resists rising.

If, then, we can find with how much less force the heavy body can be drawn up on line *bd* than on line *ba,* we will then have found with how much greater force the same heavy body descends on line *ab* than on line *bd.* And, similarly, if we can find how much greater

---

[34]On this argument, see: **Le Mecaniche** edited by Romano, Gatto, op. cit. p. CXXX–CXXXIII.

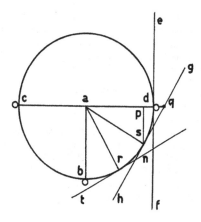

Fig. 3.2

force is needed to draw the body upward on line *bd* than on *be*, we will then have found with how much greater force the body will descend on *bd* than on *be*.

But we shall know how much less force is required to draw the body upward on *bd* than on *be* as soon as we find out how much greater will be the weight of that body on the [inclined] plane along *bd* than on the plane along *be*. Let us proceed then to investigate this weight.»[35]

Even though the last phrase can induce one to think of the medieval tradition of Jordanus of Nemore and his concept of "gravitas secundum situm", actually, as Stillman Drake says,

«… the Jordanus tradition appears to have had little if any influence on Galileo. The reason, as with Guido Ubaldo, was probably the sheer impact of the Archimedean revival. It is often said that Galileo's analysis of the inclined plane is similar to that of Jordanus, though in fact they have nothing in common beyond the correct result.»[36]

As a matter of fact, the treatment of Galileo seems to be partly inspired by that of Pappus[37] (with the recourse to the two similar right-angled triangles) and also by the Pseudo Aristotle, with the circle and the balance:

«Consider a balance *cd* (see Fig. 3.2), with center *a*, having at point *c* a weight equal to another weight at point *d*. Now, if we suppose that line *ad* moves toward *b*, pivoting about the fixed point *a*, then the descent of the body, at the initial point *d*, will be as if on line *ef*. Therefore, the descent of the body on line *ef* will be a consequence of the weight of the body at point *d*. Again, when the body is at *s*, its descent at the initial point *s* will be as if on line *gh*; and hence the motion of the body on *gh* will be a consequence of the weight that the body has at point *s*. And again, at the time when the body is at point *r*, its descent

---

[35]E. N. I, pp 296–298, comprising the following excerpt (Translation from Galileo Galilei **On Motion and On Mechanics**—edited and translated by I.E. Drabkin and Stillman Drake—The University of Wisconsin Press, 1960, pp 63–65).

[36]See: **Mechanics in the Sixteenth-Century Italy** by Stillman Drake and I.E. Drabkin—Madison, 1969, p. 55.

[37]**A Source Book in Greek Science**, op. cit. p. 195.

at the initial point r will be as if on line tn; hence the body will move on line tn in consequence of the weight that it has at point r. If, then, we can show that the body is less heavy at point s than at point d, clearly its motion on line gh will be slower than on ef. And if, again, we can show that the body at r is still less heavy than at point s, clearly the motion on line nt will be slower than on gh.

Now it is clear that the body exerts less force at point r than at point s, and less at s than at d. For the weight at point d just balances the weight at point c, since the distances ca and ad are equal. But the weight at point s does not balance that at c. For if a line is drawn from point s perpendicular to cd, the weight at s, as compared with the weight at c, is as if it were suspended from p. But a weight at p exerts less force than the [equal] weight at c, since the distance pa is less than distance ac. Similarly, a weight at r exerts less force than an [equal] weight at s: this will likewise become clear if we draw a perpendicular from r to ad, for this perpendicular will intersect ad between points a and p. It is obvious, then, that the body will descend on line ef with greater force than on line gh, and on gh with greater force than on nt.

But with *how much* greater force it moves on ef than on gh will be made clear as follows, viz., by extending line ad beyond the circle, to intersect line gh at point q.

Now since the body descends on line ef more readily than on gh in the same ratio as the body is heavier at point d than at point s, and since it is heavier at d than at s in proportion as line da is longer than ap, it follows that the body will descend on line ef more readily than on gh in proportion as line da is longer than pa. Therefore the speed on ef will bear to the speed on gh the same ratio as line da to line pa. And as da is to pa, so is qs to sp, i.e., the length of the oblique descent to the length of the vertical drop. And it is clear that the same weight can be drawn up an inclined plane with less force than vertically, in proportion as the vertical ascent is smaller than the oblique.

Consequently, the same heavy body will descend vertically with greater force than on an inclined plane in proportion as the length of the descent on the incline is greater than the vertical fall.»[38]

At the end of the "demonstration" which, for scrupulousness, we have chosen to report integrally, Galileo warns that, in order that all works, it is obviously necessary that there are not frictions of any kinds and the surfaces are completely rigid. Also, all must take place without any influence of the medium.

The hydrostatic model of Archimedean origin is, for the moment, put aside and he turns to the balance. Always and anyway, up to now, Galileo has applied principles of the statics for solving essentially dynamical problems: the dynamics was always introduced by supposing the velocity being proportional to the (specific) weight of the moveable. Also the motion in the plane is solved, in the rest of the chapter, resorting to the balance.

By making always use of geometrical arguments, he also obtains that, given different inclined planes of the same height, but of different length, the ratio of the velocities of the same moveable slipping on them is equal to the inverse ratio of the respective lengths.

The big stumbling block which lies on the way of Galileo is the acceleration due to gravity.

---

[38]See: [Aristotele] **Meccanica** a cura di Maria Fernanda Ferrini—Bompiani, 2010—p. 169.

The "natural" motion he deals with is due to gravity, but according to him it always and anyhow results to be a uniform motion.

For now, he does not succeed in getting rid of the Aristotelian tenet «velocity proportional to force» and this is the first cause of the failure of the **De Motu**.

It is also of great importance, in the general economy of the Treatise, the poverty of the specific language, that is the lack of true definitions (with related lexicon) of the magnitudes which are used. This serious infantile illness of the Galileian mechanics is thoroughly analysed in the irreplaceable work of Paolo Galluzzi which follows the evolution of the Galileian language in which, starting from the work **Le Mechaniche** (subsequent to the **De Motu**), the term "momento"[39] will be fundamental. The impression one receives in reading this chapter of the Treatise is that Galileo had certainly aimed to solve with it the problems of the ("natural") motion definitively, also introducing the new paradigms of the balance and thinking about an impossible synthesis with the hydrostatic model.

Unfortunately, as Galluzzi remarks, «… of the ambitious project only a heap of rubble remained.».[40] Nonetheless, one has to do with very important pages for the origin of the (future) Galileian mechanics. In fact, it appears, for the first time, what we could call a proto-principle of inertia: on a horizontal plane, a material sphere can be moved by a minimal force (i.e., it is the same as to say that a sphere still on a plane is not retained by any force). He takes care to tell us that, since we cannot be free from any external influence, on the case someone wanted to try the experience would remain disappointed, since he should exert a non-negligible force to move the sphere.

In addition, he informs us that, the Earth being spherical, an infinite plane parallel to the horizon really does not exist, but we have a spherical surface of a greatest radius having its centre at the centre of the Earth. As it is known, this consideration has caused that volumes were written on the so called "principle of circular inertia", on which we shall come back in due time.[41]

In any case, in this chapter itself, Galileo has invoked the authority of Archimedes in considering as parallel lines (rather than convergent in the centre of the Earth) the lines perpendicular to the bar of the balance. Therefore, in the neighbourhood of a point, the tangent plane can substitute the spherical surface.

3.4.2 The Circular Motion

When we speak of circular motion, nowadays, we intend to refer, without further specifications, to the motion of a mass point on a circumference. In the case

---

[39]Paolo Galluzzi: **Momento—Studi Galileiani**—Edizioni dell'Ateneo & Bizzarri, Roma 1979.

[40]Ibidem, p. 193.

[41]We shall also see that the "proto-inertia principle", which here seems to be a mere consequence of the demonstration regarding the inclined planes, will assume during the time an essential role for Galileo.

of Galileo, instead one has to do with the motion of a sphere or around a point (which may be the centre of figure or the centre of gravity) or around a diametral axis.

Of course, the single points of the sphere will always go along a circumference but the important point to which one must refer is not, in any case, the centre of the circumference, but the centre of the world.

The problem that one deals with is that of deciding if and in what cases the motion is natural or forced or neither of them. As a premise, Galileo states again the criterion for distinguishing the two kinds of motion:

«Motus itaque naturalis est dum mobilia, incedendo, ad loca propria accedunt; violentus vero est dum mobilia, quae moventur, a proprio loco recedunt.»[42]

Therefore, the motion is natural when the moveable goes toward the centre of the world and, on the contrary, forced (or violent) when it goes away from it.

Let us see the possible cases for the rotation of the sphere around its centre. The centre of rotation may coincide with the centre of the world or not. In the first case, if the sphere is homogeneous, the centre of gravity and the centre of figure coincide and coincide with the centre of rotation, if the sphere is not homogeneous can rotate around the centre of gravity or around the centre of figure.

Also in the case where the centre of gravity does not coincide with the centre of the world, we are always in the situation in which it neither goes away nor goes near the centre of the world itself (in fact, every point goes along a circumference around it). Therefore, when the sphere rotates around the centre of the world, the motion is neither natural nor violent.

In the second case, one considers the rotation of a sphere around a diametral axis out of the centre of the world. If the sphere is homogeneous the axis passes through the centre of gravity and then the centre of gravity itself, during the rotation, neither goes away nor goes near the centre of the world.

Thus we are in the condition of the first case. Things change, on the contrary, if the sphere is not homogeneous. Now the rotation happens about the centre of figure and the centre of gravity goes away and near the centre of the world during the motion. Therefore, the motion will be alternately natural and violent.

Galileo also hints at the old problem of the slowing down of the motion of rotation, promising to deal with it later on. One can conclude that there is nothing new with respect to what said in the Dialogue.

3.4.3 The Motion of the Projectiles

The projectiles, according to the ancient nomenclature, are the bodies which are thrown by persons or by mechanisms without being directly followed in their

---

[42]E. N. I, p. 304 (17–19). «Now we have natural motion when bodies, as they move, approach their natural places, and forced motion when the bodies that move recede from their natural place».

motion by the entity which gave origin to their motion. A stone thrown in the space or an arrow shot by a bow are typical examples.

We know that, according to Aristotle, the projectiles, after having been released by the projector, were supported in their motion by the medium in which they moved through the mechanism of the antiperistasis. Therefore, the medium was absolutely necessary to the motion.

Galileo exposes the thought of Aristotle regarding this problem with a great deal of sarcasm, and opposes to it his theory (which is then that borrowed from Avempace). In his opinion, the projector impresses to the projectile a force (*virtus motiva*)[43] which deprives it of its gravity, when it is thrown upward, or of its lightness when it is pushed downward (as one can see, maybe for convenience of expression, the concept of "levitas" has reappeared).

To make clear how the *virtus motiva* can deprive the moveable of its qualities (gravity or lightness), Galileo appeals to arguments substantially "Aristotelian"; the fire deprives the iron of the cold, by introducing in it the warm, the hammer deprives the bell of the silence by introducing in it sound, etc., i.e. the *virtus motiva* produces changes of quality, although destined not to last. In fact, the *virtus motiva* so introduced in the projectile, in the long run will come to an end as the sound of the bell or the warm in the iron which will return cold.

Starting from the assumption that the violent motion must be one and continuous, contrarily to what maintained by Aristotle with his mechanism, and from the fact that it cannot last indefinitely, Galileo easily succeeds to demonstrate that the *virtus motiva* comes spontaneously to an end.

The process of extinction of the impressed force also provides him with the explanation of the acceleration of fall of a body thrown upward. In fact, for throwing a body upward, one must impress in it a force greater than the weight. This force will progressively decrease till becoming equal to the weight; at this point the motion is inverted and the body begins to go down subject to two forces, the weight (constant) and the impressed force contrary to it but continuously decreasing.

Therefore, the velocity (continuously decreasing) due to the impressed force is subtracted from the constant velocity of fall, proportional to the specific weight of the body. The result is that, until the impressed force is completely extinguished, the motion will be accelerated until reaching the velocity (which from now on will remain constant) that is due to that body in relation to its specific weight.

As regards the point of inversion of the motion, the moment of rest assumed by Aristotle is excluded by showing that the vanishing impressed force cannot remain equal to the gravity for a finite interval of time.

Then by further exploiting the same ingredients (self-vanishing impressed force, velocity of the natural motion proportional to the specific weight), Galileo tries to answer the other questions on the motion of the projectiles he had listed in the programme we have seen at the beginning. The result is obviously

---

[43]E. N. I, p. 309 (31–32).

disappointing, given the starting point. It must be remarked that Galileo is still convinced that the trajectory of a projectile consists of three tracts, two rectilinear segments connected by a curved tract, even if he does not study in depth the subject.

3.5 Conclusions

We cannot resist the temptation, following Galileo's way, to establish a proportion: we could say that the interest the **De Motu** excites, and has excited, is in inverse proportion with the reached results.

M. Clavelin says:

> «Pourtant l'intérêt du **De Motu** est certain. Traitant simultanément du mouvement naturel et du mouvement des projectiles, c'est bien une dynamique générale que Galilée s'efforce déjà de construire. En même temps le recours systematique à Archimède exprime clairement la volonté de substituer à la méthode historique et dialectique des philosophes une méthode directe et quantitative. Oeuvre polémique et libéretrice, le **De Motu** traduit avec fidélité le premier état de la science galiléenne.»[44]

And, further on,

> «Travail de critique et de synthése, le **De Motu** est donc loin d'être sans intérêt. Le retour, quelque quarante-cinq ans plus tard, dans la premiére journée des **Discours**, de certains de ses arguments suffirait à en témoigner …».[45]

We can but agree with Clavelin. The **De Motu** is a polemic and liberating work: Galileo tries to get out of the choking burden of the Aristotelian natural philosophy, even though he has not yet at his disposal the necessary mathematic-physical tools (which he thinks to have found in the Archimedean hydrostatics). The polemic against Aristotle is rather heated, arriving till derision. The opinions of the "philosopher" are defined ridiculous and he himself accused of being ignorant even about elementary mathematics.

If it is clear who was the "enemy", not everyone had the same opinion about the identity of the "friends". Clearly, the problem is that of the "sources" and about this the opinions are diverging.

The fact that we have placed Moody's opinion in epigraph form in Sect. 3.3, immediately characterizes us as not being a member of the group of the upholders of the medieval tradition (both Mertonian and Parisian).

In this direction it seems to move, actually, also a fair part of the experts, once cooled down in the last decades the infatuation for the theses of Duhem.

Galileo never quotes, not even hastily, neither the Mertonians nor the Parisians, but the champions of the "medieval" thesis hold that this is not an evidence he did

---

[44]Maurice Clavelin: **La philosophie naturelle de Galilée**—Albin Michel, 1996 (première édition, 1968), p. 130.

[45]Ibidem, p. 141.

not know them. It is true that Galileo, in his maturity, was not generous with quotations, but the young man struggling with his first work about the motion crowds his notes (the so called Memoranda) with names.

In Chap. X of the Treatise, by holding his thesis on the existence of the void and the non-instantaneousness of the motion in it, he explicitly involves Duns Scoto, Thomas Aquinas and Philoponus as authors in disagreement with Aristotle:

«... Scotus, D. Thomas, Philoponus et alii nonnulli contrariam Aristoteli teneant sententiam, attamen veritatem fide potius quam vera demonstratione, aut quod Aristoteli responderint, sunt consecuti.»[46]

He refers to the same authors, with the addition of Avempace and Avicenna, in the Memoranda as well,[47] again in the case of the motion in the void. The *vis*, or *virtus*, impressed comes from Avempace, through Borri. In fact, Borri's text is the only work of a contemporary author which Galileo refers to[48] and just in Borri's book the theory of Avempace is extensively exposed, together with its confutation by Averroes (Avempace had been the master of Averroes).

Before referring to significant passages of Borri's book, we remark that first he states a list of two pages of «*Veterum philosophorum nomina, quorum sententiae in hoc libro aut recipiuntur, aut exploduntur*».

The list begins with Alcinous and ends with Zeno and, besides Avempace, also contains Philoponus. Borri does not consider the philosophers following Averroes (1126–1198) noteworthy and, therefore, the medieval tradition of Mertonians and Parisians is completely ignored: it is as it never existed. Thus, the natural philosophy (to be both received and rejected) ends with Averroes. Then, this is also the philosophy that must be fought: the natural philosophy of Aristotle, revisited by Averroes.

Borri's book is divided into three parts and the explanation of the theory of Avempace and its refutation take two chapters (one in the second part and one in the third). In fact, Chap. XXVII of the second part (*Quae Averroès contra doctissimum Avempacem praeceptorem pro Aristotele scripsit*)[49] and Chap. XIX of the third part (*Avempacis argumentum ad aequam lancem expenditur, quod quanto Physicorum ab Avempace texitur*)[50] are completely devoted to the theory of Avempace, of which an impartial exposition (*aequa lance*) and a refutation are promised.

In our opinion, this is the main source, for Galileo, of the impressed force. With regard to the rare use of the term "impetus", it seems to us that it should absolutely

---

[46]E. N. I, p. 284 (7–10). «And though Scotus, Saint Thomas, Philoponus, and some others hold a view opposed to Aristotle's, they arrive at the truth by belief rather than real proof or by refuting Aristotle».

[47]Ibidem, p. 410 (21–26).

[48]Besides having almost certainly attended the lectures of Borri at Pisa, Galileo also owned a copy of the book (see A. Favaro: **La libreria di Galileo Galilei**—Bullettino di Bibliografia e di Storia della Scienze Matematiche e Fisiche—tomo XIX—p. 268 (1886)).

[49]Borri, op. cit. p. 124.

[50]Ibidem, p. 251.

not referred to a loan from the theory of Buridan and his followers. It is obviously superfluous to insist on the dependence on Archimedes for all the "hydrostatic" part. Further, it is also our opinion that the monumental work of Francesco Bonamici[51] (professor at Pisa when he was student), even if owned by Galileo,[52] resulted irrelevant for the preparation of the young Galileo.

At last, it remains the "question Benedetti".

As we have reminded in the second chapter, the work of Giovanni Battista Benedetti was "rediscovered" by Guglielmo Libri in 1840 and further on by Giovanni Vailati (1898) and, in the same years, also illustrated by Raffaello Caverni. Vailati and Caverni had drawn the work of Benedetti near the juvenile work of Galileo, but it is due to the international repute of the former if Benedetti received the recognition of direct forerunner of Galileo in the international literature.[53]

Vailati says, with regard to the independence of the velocity of the heavy bodies from their (absolute) weight:

«The first statement of this fundamental law, at least for that part which regards the comparison between heavy bodies constituted by different quantities of the same matter, is unquestionably due to Benedetti, and it is important to remark that the reasons he adduces for proving it, perfectly coincide with those by which Galileo, a quarter of a century later, asserted to have been driven to the same discovery».[54]

This is the basis on which what we should call a circumstantial trial has been instituted. Another "circumstantial evidence", which we have already reminded in the second chapter, is given by the fact that Jacopo Mazzoni in his work quotes many times the **Diversarum Speculationum** of Benedetti as a text of reference for the physics and, then, could have advised that text to Galileo (in the Pisan period).

On this supposition, as we have seen, Caverni had constructed a story.

The "circumstantial evidences" contrary to the thesis formulated are "in absentia". Galileo has never written the name of Benedetti neither in the published works nor in the manuscripts come to us, and the works of Benedetti have not been found among the books whose property has been attributed to Galileo (whereas the works of Tartaglia have been found).[55]

A careful study of Benedetti's work, like that performed by Enrico Giusti,[56] also allows us to control the difference between Benedetti and the young Galileo.

The result is that we agree with R. Fredette when he says that, until someone will succeed to effectively demonstrate the contrary, he remains convinced that Galileo did never read the works of Benedetti.[57]

---

[51]Francisci Bonamici Florentini: **De Motu Libri X**—Florentiae, MDXCI.

[52]A. Favaro: **La libreria di Galileo Galilei**, loc. cit. p. 242.

[53]It is enough, for all, to refer to A. Koyrè: **Études Galiléennes**, op. cit..

[54]Giovanni Vailati: **Scritti**, op. cit. vol. II, p. 151.

[55]A. Favaro: **La libreria di Galileo Galilei**, loc. cit. p. 268.

[56]See: E. Giusti—**Le speculazioni giovanili «De Motu»**—di G. B. Benedetti—loc. cit..

[57]R. Fredette: **Galileo's De Motu Antiquiora—Notes for a reappraisal**—loc. cit. p. 180.

That being stated, it remains to reaffirm why this little juvenile work must be considered important. Obviously it is, because, like in many cases of this kind, it allows one to know what were at the beginning the Galilean problems, but also and, we should say, above all, it is the first systematic attempt to reject the Aristotelian theory of motion and construct a new one. The fact that the result, at the end, is constituted by "rubble", as commented by Galluzzi, does not cause us to underestimate its importance.

From the study of the work of Archimedes, Galileo has had clear that, whereas for the statics of that time one was in possession of a mathematized and then rigorous theory, in the case of the dynamics one had only a series of statements devoid of real proofs. In fact, some of Aristotelian "demonstrations" are by him easily invalidated.

But he has not the tools for building the new theory and therefore he will fail by trying to construct a dynamics of uniform motion where the acceleration is only an "accident".

3.6 Additional Considerations

Since, according to the scheme decided for this book, we resolved to study the development of the Galilean theory of motion, this bring us to isolate from the context all that regards the law and the principles relevant to the motion. In the case of **De Motu**, obviously, this did not require any particular arrangement, since the work was entirely devoted to the motion.

Since a continuity exists, not only temporal, but also of cultural evolution, with the work successively left in manuscript form as well (i.e. **Le Mecaniche**), let us try to understand, before passing to the works of the maturity, how Galileo is constructing his mechanics bit by bit.

Certainly one can say that, from a certain point on, his interest in mechanics should coexist with other interests which from one hand widened his field of action, but from the other hand slackened the study of mechanics.

Then let us see how things happened.

The 26 September of 1592 Galileo was appointed professor of Mathematics at the University of Padua. Most probably, already in his first academic year,[58] he gave a course of lectures devoted to those which were named «Mechanical Questions of Aristotle» (i.e. that work on the machines which in the antiquity was attributed to Aristotle). Handwritten copies of this course were found starting from 1898, therefore too late for the second volume of the National Edition (1891). From then on they are known as the "short version" of **Le Mecaniche**.

---

[58]For this and for the dating of the two versions of **Le Mecaniche**, we refer to the work of Romano Gatto who has given a critical edition of both versions, equipped with an exhaustive introductory essay. See: Galileo Galilei—**Le Mecaniche**, op. cit. pp LIII–LXX.

Subsequently, in the academic year 1598–1599, Galileo gave again a course of lectures on that subject (this time with a greater widening) and from the leftover handwritten copies the so-called "long version" was drawn (that published in the second volume of the N. E.). Therefore, between the two versions, there is an interval of six years. Even a French translation (that we could define a "free translation") of the "long version" was published by Marin Mersenne (of the order of Minims) in 1634.[59]

One must not be astonished of this fact, since at that time many handwritten copies of **Le Mecaniche** circulated in Europe. In Italy, instead, the work was published posthumous edited by Luca Danesi in 1649.[60]

We are dwelling on this information, not for a mere documentary attitude, but for giving proof that the interest of Galileo in the theory of motion had not ceased (waiting for better times) with the end of the Pisan period. In fact, by comparing the two versions of **Le Mechaniche**, one clearly perceives that a progress in the theoretical formulation has occurred.

Now, Galileo lists some definitions and hypotheses (*supposisizioni*) and, among the definitions, he gives that of moment:

«*Moment* is the tendency to move downward caused not so much by the heaviness of the movable body as by the arrangement which different heavy bodies have among themselves. It is through such *moment* that a less heavy body will often be seen to counterbalance some other of greater heaviness, as in the steelyard a little counterweight is seen to raise a very heavy weight, not by excess of heaviness, but rather by its distance from the suspension of the steelyard. This, combined with the heaviness of the lesser weight, increases its *moment* and impetus to go downward, with which it may exceed the *moment* of the other, heavier weight. Thus *moment* is that impetus to go downward composed of heaviness, position, and of anything else by which this tendency may be caused.»[61]

It is the definition of "mechanical moment" (a little longer than that in the modern scientific language!) which, as Paolo Galluzzi remarks, appears for the first time with this meaning.[62]

In the course of the treatment of the motion along an inclined plane (which belongs to the chapter on the screw) the "moment to go downward" happens many times. These are signals which testify that the study of the motion had continued, maybe, in a certain sense, isolatedly in those years; in fact, there are no written traces regarding the period included between the two versions of **Le Mecaniche**.

Also the exchange of letters with Guidobaldo del Monte regards essentially what Guidobaldo was writing (a work on the perspective and one on the

---

[59]**Les mechaniques de Galilée** Mathematicien & Ingenieur du Duc de Florence …—Traduit de l'Italien par L. P. M. M.—A Paris, Chez Henry Guenon—MDCXXXIV. A recent critical edition of this work exists, edited by Bernard Rochot, PUF, 1966.

[60]**Della Scienza della Meccanica e delle utilità che si traggono dagl'istromenti di quella**. Opera cavata dai manoscritti dell'eccellentissimo matematico Galileo Galilei dal cav. Luca Danesi—Ravenna, stamp. Camerali, 1649.

[61]Galileo Galilei: **Le Mecaniche**, op. cit. pp 48–49. English translation by Stillman Drake in **On Motion and on Mechanics**, op. cit., p. 151.

[62]See: Paolo Galluzzi: **Momento**, op. cit. pp 199–227. Galluzzi emphasizes that in the short version the term appears in the meaning of a small quantity.

Archimedean screw which will be published posthumous). In any case, also in this occasion it is repeated the argument regarding a sphere, still on a plane, which can be displaced by minimal force—i.e. the "proto-inertial" principle.

However, this principle is not caught by Mersenne who, in two additions to the chapter on the screw (III Addition and IV Addition), insists on the fact that one must have to do with a homogeneous sphere tangent in one point to the plane.[63] Galileo, instead, has drawn that «From this we may take the following conclusion as an indubitable axiom. The heavy bodies all external and adventitious impediments being removed, can be moved in the plane of the horizon by any minimum force.»[64]

It is just the case to say we are on another plane at all.

We can conclude with Clavelin:

«Le **De Motu** et **Le Mecaniche** permettent donc de dresser un tableau assez complet des premiers développements de la pensée galiléenne. D'abord simplement commentée, la tradition est progressivement dominée, puis reconstruite; dans la précision et la trasformation des concepts anciens c'est la science nouvelle qui déja s'announce. De l'une à l'autre oeuvre, sourtant, Galilée prende plus en plus clairement conscience de ce que signifie la mathematisation de la philosophie naturelle.»[65]

Bibliography

We quote here only books and papers not mentioned in the footnotes of the chapter.

A. Carugo, A.C. Crombie, *The Jesuits and Galileo's ideas of science and of nature*. Annali dell'Istituto e Museo di Storia della Scienza, Firenze **VIII–2**, 3–68 (1983)

I.E. Drabkin, *A note on Galileo's De Motu*. Isis **51**, 271–277 (1960)

S. Drake, *The earliest version of Galileo's mechanics*. Osiris **13**, 262–290 (1958)

S. Drake, *On the probable order of Galileo's notes on motion*. Physis **14**, 55–68 (1972)

S. Drake, *Galileo's notes on motion*. Istituto e Museo di Storia della Scienza, Firenze. Monografia N. 3 (1979)

R. Giacomelli, *Galileo Galilei giovane e il suo "De Motu"*. Domus Galilaeana, Pisa (1949)

L. Winifred Wisan, *The new science of motion: a study of Galileo's De Motu locali*. Arch. Hist. Exact Sci. **13**, 103–306 (1974)

---

[63]**Les Mechaniques de Galilée** op. cit. pp 58–62.

[64]Galileo Galilei: **Le Mecaniche**, op. cit. p. 68.

[65]Maurice Clavelin: **La philosophie naturelle de Galilée**, op. cit. p. 177.

Chapter 4
The Inertia Principle

According to the majority of the historians of physics, the inertia principle marks the beginning of modern physics. A prerequisite for its enunciation is not only the admission of the possible existence of the void, but also of the possibility of motion in it.

Paradoxically, as observed by Edward Grant, in this regard we could credit to Aristotle the first enunciation of the inertia principle.

In fact, in the already quoted fourth book of **Physics**, in one of the passages regarding the void, he says:

> «Further, no one could say why a thing once set in motion should stop anywhere; for why should it stop here rather than there? So that a thing will either be at rest or must be moved ad infinitum, unless something more powerful get in its way.» (**Physics**, IV8, 215a, 19–21).

In the Middle Ages, as Grant remarks, this part of Aristotle's discourse has not caused remarks and the problem of the motion in the void, in opposition to Aristotle, as we have already seen, has attracted attention in consequence of a revived interest in the ideas of Philoponus and Avempace.

We resolved to study the origin (and the development) of the "results" which are traditionally credited to Galileo. In the traditional vulgate, the inertia principle is one of them and since in the foundations of classical mechanics, i.e. in Newton's **Principia**, it takes the first place, it is natural that also by us it is taken into consideration first and with reference to the Newtonian formulation.

> As I became interested in the history of science and in the work of Galileo in particular, I came to realize that the origin of the law of inertia, and Galileo's role in it, involved questions more intricate than is generally supposed, being still subject of study and debate today. And there are few questions more fascinating in the whole history of physics. One need only try to conceive of a science of physics without the concept of inertia in order to perceive that the introduction of this fundamental notion must have produced a revolution in physical thought as profound as any that have occurred since.
>
> Stillman Drake: **Galileo and the law of inertia**—American Journal of Physics, 37, p. 602, 1964.

© Springer International Publishing Switzerland 2016
D. Boccaletti, *Galileo and the Equations of Motion*,
DOI 10.1007/978-3-319-20134-4_4

4.1 The Lex I of Newton

The so-called "Inertia Principle", or first Newton's law, is usually exposed in the elementary textbooks of physics, also nowadays, by substantially repeating the statement of the Lex I of Newton's **Principia**. The definitions Newton put before the statement of his three laws, and the three laws themselves, have been subjected to a close criticism from the point of view of their logical coherence and some of them, starting from Mach,[1] regarded as tautologies.

We shall not deal with this question and the definition of an inertial frame of reference, neither with the various enunciations in axiomatic form which in the course of time have been proposed for making up for the inconsistencies of the Newtonian formalization.

Our aim is that of investigating if the attribution to Galileo of the paternity of the inertia principle, in the above-quoted "Newtonian" form, is justified or not.

First of all, we warn the reader that the question is not of the simplest ones, as it might appear at first sight, since it is still a battleground for the specialists. The subject has been discussed by generations of scholars (for the last century, it is sufficient to remind the names of Alexandre Koyré and Stillman Drake, occupying opposite positions) and, as we shall see, the reached conclusions are not universally shared.

As an introduction, in order to make first clear what must be the term of reference, we report the enunciation of the three Newton's laws integrally. As it is known, Newton's **Principia** have had three editions during the lifetime of the author: the first in 1687, the second in 1713, the third in 1726. In the following we report the three laws from the first edition.[2]

> Lex I
>   «Corpus omne perseverare in statu suo quiescendi vel movendi uniformiter in directum, nisi quatenus a viribus impressis cogitur statum illum mutare»[3]
> Lex II
>   «Mutationem motus proportionalem esse vi motrici impressae, & fieri secundum lineam rectam qua vis illa imprimitur.»[4]
> Lex III
>   «Actioni contrariam semper & aequalem esse reactionem: sive corporum duorum actiones in se mutuo semper esse aequales & in partes contrarias dirigi.»[5]

---

[1]See: E. Mach, op. cit..

[2]**Philosophiae Naturalis Principia Mathematica**—Autore Is. Newton—Londini—Anno MDCLXXXVII—pp 12–13.

[3]«Every body perseveres in its state of being at rest or of moving uniformly straightforward, except insofar as it is compelled to change its state by forces impressed». The translation is from: Isaac Newton—**The Principia**—A new translation by I. Bernard Cohen and Anne Whitman—University of California Press, 1999, p. 416.

[4]«A change in motion is proportional to the motive force impressed and takes place along the straight line in which that force is impressed». Op. cit., ibidem.

[5]«To any action there is always an opposite and equal reaction; in other words, the actions of two bodies upon each other are always equal and always opposite in direction». Op. cit., p. 417.

The enunciation of the three laws is remained unchanged from the first to the third edition,[6] except for a slight correction of purely syntactic nature in the text of the **Lex I**.

Let us see it explicitly.

The text remains unchanged till the comma; after the comma, it states:

«…, nisi quatenus illud a viribus impressis cogitur statum suum mutare.»

As it can be seen, in the four decades elapsed no change of mind has occurred. Only a little variation in the "Latin". Therefore this will be, for us, the enunciation of reference. We do not consider what precedes a useless meticulousness.

As we shall have a chance of seeing later on, in the case of Galileo, the interpretation of the development of his thought from the period of the teaching at Pisa to the printing of the **Discourses** is crucial for the objective we set ourselves.

We further add to what we have already quoted, that in the Scholium which follows Corollary VI, both in the first edition (p. 20) and in the third (p. 21), Newton recognizes, so to say, the priority of Galileo:

«Per leges duas primas & corollaria duo prima adinvenit Galilaeus descensum gravium esse in duplicata ratione temporis, & motum projectilium fieri in Parabola, conspirante experientia, nisi quatenus motus illi per aeris resistentiam aliquantulum retardantur.»[7]

As it is universally known, Newton was anything but easily open to recognize the "priority" of anybody else. Therefore, his admission, although historically incorrect (in the sense that he exchanges the logical ex post reconstruction for the scientific process effectively occurred), has a great significance.

It was not the customary eulogy of the great men of the antiquity with the recognition due to the predecessors (a dwarf on the shoulders of giants), since Galileo was died only forty five years before.

It was rather an assent to an opinion shared in "the community of experts". Even if Galileo never expressed the inertia principle in the formulation of Newton's Lex I, however he had laid the foundations on which it could be constructed.

4.2 The Inertia in the De Motu

We have already dealt with the fourteenth chapter (or first of the second book) of **De Motu** when speaking of the motion along an inclined plane. In that occasion we also hinted at an inertia principle, so to say yet in a larval stage, which appeared inserted in a certain evidence in order to demonstrate that a moveable on a plane surface could be moved by a minimal force.

---

[6]**Philosophiae Naturalis Principia Mathematica**—Auctore Isaaco Newtono—Editio tertia aucta & emendata, Londini, MDCCXXVI.

[7]«By means of the first two laws and the first two corollaries Galileo found that the descent of heavy bodies is in the squared ratio of the time and that the motion of projectiles occurs in a parabola, as experiment confirms, except insofar as these motions are somewhat retarded by the resistance of the air». Op. cit., p. 424.

Before to examine into details, directly from the source, what Galileo wrote, we would put a consideration before. It is a common practice, on the part of the historians of science, to compare the statements of Galileo with analogous statements of contemporaries or predecessors (see, for instance, the comparison with Benedetti of which we have already treated), chosen, so it seems to us, by taking them out of the context.

At any time, certain problems concern not a single person but a circle, even if narrow, of scholars. For instance, already a century and half before Galileo, Nicholas of Cusa in his **De staticis Experimentis** (1450), also a little work in form of a dialogue, tackled the problems (to avoid any misunderstanding, those related to the specific weight) which Galileo deals with starting from the Dialogue and then in the Treatise of **De Motu**. Cusa himself in an other dialogue, **De Ludo Globi** (1464), deals with the possible motion of a sphere and, regarding the motion on a plane:

«… Ideo sphaera in plana et aequali superficie se semper aequaliter habens, semel mota, semper moveretur. Forma igitur rotunditatis ad perpetuitatem motus est aptissima. Cui si motus advenit naturaliter, numquam cessabit».[8]

With this we want to say that it is not the case of considering Nicholas of Cusa a "predecessor" to be compared with Galileo, both for the problems of the specific weight and the motion of a moveable on a plane, but that, from a certain point on, some problems become object of common interest within a certain community of experts.

Therefore, and this is for accounting for the introduction of those excerpts which can be read as a first intuition of the inertia principle, what we want to try of understanding is how, from a yet unsolved whole of problems dealt with by different persons, a solution destined to converge in a new physical law can emerge.

We have already reminded in Sect. 3.3.1 the contribution of Cardano to the treatment of the so-called "problem of Pappus", i.e. the determination of the force necessary to move a body along an inclined plane. Let us remind, now, for completeness, that also Guidobaldo Del Monte, in his first work **Mechanicorum Liber** (Pesaro, 1577), at the end deals with the problem of Pappus, but does not modify its solution.

The phrase «With the wedge, it seems clear that one can never move a given weight by means of a given power, because a given power cannot move a given weight along an inclined plane».[9] makes Stillman Drake say: «This succinctly illustrates the distance that still separated the most systematic sixteenth-century treatment of mechanics from the inertial concept».[10]

---

[8]See: **De Ludo Globi**—Editio Parisiensis (1514) Vol. I, fol. 154v. «And so, a sphere that behaved always in the same way, on a flat and even surface, would always be moved, once it began to be moved. Therefore, the form of roundness is the form that is most suitable for the perpetuity of motion. If motion is natural to this form, then the motion will never cease.» The English translation (by Jasper Hopkins) is taken from **The Bowling—Game,** I, 21—Cusan—Portal.

[9]From **Mechanics in Sixteenth-Century Italy**—Selection from Tartaglia, Benedetti, Guido Ubaldo & Galileo—Translated & annotated by Stillman Drake & I.E. Drabkin—University of Wisconsin Press, 1969, p. 327.

[10]Ibidem.

At the end of the "geometric" demonstration regarding the motion along an inclined plane we have integrally reported in the third chapter, and after having repeated the usual reservation regarding the experimental apparatus, Galileo concludes:

«If everything is arranged in this way, then any body on a plane parallel to the horizon will be moved by the very smallest force, indeed, by a force less than any given force. Since this seems quite hard to believe, it will be proved by the following demonstration».[11]

After having again recoursed to the balance for the demonstration, he finally asserts:

«But there is more. A body subject to no external resistance on a plane sloping no matter how litle below the horizon will move down [the plane] in natural motion, without the application of any external force. This can be seen in the case of water. And the same body on a plane sloping upward, no matter how little, above the horizon, does not move up [the plane] except by force. And so the conclusion remains that on the horizontal plane itself the motion of the body is neither natural nor forced. But if its motion is not forced motion, then it can be made to move by the smallest of all possible forces.»[12]

For securing himself against an immediate denial on part of anybody who tried to perform the experiment, he insists:

«And our demonstrations, as we also said above, must be understood of bodies free from all external resistance. But since it is perhaps impossible to find such bodies in the realm of matter, one who performs an experiment on the subject should not be surprised if the experiment fails, that is, if a large sphere, even though it is on a horizontal plane, cannot be moved with a minimal force. For in addition to the causes already mentioned, there is also this one that a plane cannot actually be parallel to the horizon, since the surface of the earth is spherical, and a plane cannot be parallel to such a surface. Hence, since [in the aforesaid experiment] the plane touches the sphere in only one point, if we move away from that point, we shall have to be moving up. There is good reason, therefore, why it will not be possible to move the sphere from that point with an arbitrarily minimal force.»[13]

It seems to us that two things must be pointed out:

First—in the first quoted excerpt, Galileo starts with "But there is more" (amplius, in Latin). This means that he does not want to limit himself to describe what happens on the different inclined planes, but that he wants to deduce from this a consequence having a general validity.

Second—at the end of the second excerpt, he delimits the generality of the obtained result. Therefore, we are in presence of a "restricted" inertia principle.

---

[11]E. N. I, p. 299 (2–5) (Drabkin pp. 65–66).

[12]Ibidem.

[13]E. N. I, pp 300–301. (Drabkin p. 68).

4.3 Inertia in Le Mechaniche and in Dimostrazioni Intorno Alle Macchie Solari E Loro Accidenti

We have anticipated, at the end of Chap. 3, that also in **Le Mechaniche** (long version) Galileo gives a version of what we have, for now, called restricted inertia principle. At the beginning of the chapter regarding the screw, "among the most beautiful and useful contrivance", he passes immediately to expose what, at the end, he will assume as "an indubitable axiom":

> «There can be no doubt that the constitution of nature with respect to the movements of heavy bodies is such that any body which retains heaviness within itself has a propensity, when free, to move toward the center; and not only by a straight perpendicular line, but also, when it cannot do otherwise, along any other line which, having some tilt toward the center, goes downward little by little. And thus we see, for instance, that water from some high place not only drops perpendicularly downward, but also runs about the surface of the earth on lines that are inclined, though but very little. This is seen in the course of rivers whose waters, though the bed is very little slanted, run freely dropping downward; which same effect, just as it is perceived in fluid bodies, appears also in hard solids, provided that their shapes and other external and accidental impediments do not prevent it.

> So that if we have a surface that is very smooth and polished, as would be that of a mirror, and a perfectly smooth and round ball of marble or glass or some such material capable of being polished, then if this ball is placed on that surface it will go moving along, provided that the surface has some little tilt, even the slightest; and it will remain still only on that surface which is most precisely leveled, and equidistant from the plane of the horizon. This, for example, might be the surface of a frozen lake or pond, upon which such a spherical body would stand still, though with a disposition to be moved by any extremely small force. For we have understood that if such a plane tilted only by a hair, the said ball would move spontaneously toward the lower part, and on the other hand it would have resistance toward the upper or rising part, nor could it be moved that way without some violence.

> Hence it is perfectly clear that on an exactly balanced surface the ball would remain indifferent and questioning between motion and rest, so that any the least force would be sufficient to move it, just as on the other hand any little resistance, such as that merely of the air that surrounds it, would be capable of holding it still. From this we may take the following conclusion as an indubitable axiom:

> That heavy bodies, all external and adventitious impediments being removed, can be moved in the plane of the horizon by any minimum force.

> But when the same heavy body must be driven upon an ascending plane, having a tendency to the contrary motion and commencing to oppose such an ascent, there will be required greater and greater violence the more elevation the said plane shall have. For example, the movable body G (Fig. 4.1) being placed on the line *AB* parallel to the horizon, it will stand there, as was said, indifferent to motion or to rest, so that it may be moved by the least force.»[14]

There are no appreciable differences from the enunciation of **De Motu**, except that it is no more reminded that the surface of the Earth is not a plane. But there is an element that, we think, has to do with the importance that Galileo attaches to this "restricted inertia principle". Rather than to come out as a consequence of the

---

[14]See: **Le Mecaniche**, op. cit. pp 68–69 (861–906) (Drake, pp 170–171).

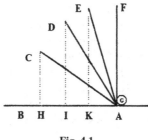

Fig. 4.1

treatment of the motion of a body along an inclined plane, now the "problem of Pappus" is put before the treatment itself and in evidence.

«This present theory was attempted also by Pappus of Alexandria in the eighth book of his **Mathematical Collections**, but in my opinion he missed the mark, being defeated by the assumption which he made when he supposed that the weight would have to be moved in the horizontal plane by a given force. This is false, no sensible force being required (neglecting accidental impediments, which are not considered by the theoretician) to move the given weight horizontally, so that it is vain thus to seek the force with which it will be moved on the inclined plane.»[15]

It is worth to be noticed, at last, that he stresses the fact that the "theoretician" leaves the "accidental impediments" out of consideration. Whereas in the past he worried of warning a possible experimenter that it would be difficult to obtain the expected result because of the aforesaid impediments, now he asserts that the "theoretician", on principle, must abstract from them. Even though Galileo has continued to deal with mechanics, and theory of motion in particular, in all the period spent in Padua, the first enunciation of the restricted inertia principle, after that of **Le Mechaniche** is in the work (this time printed) about the sunspots published in 1613.[16]

Let us summarize the antecedents. At the beginning of 1612, the Jesuit Christopher Scheiner, with the pen name of Apelle ("Apelles latens post tabulam", i.e. like the painter Apelles who hid himself behind the picture just painted waiting for the comments before revealing his identity) had published a booklet entitled **Tres Epistolas de Maculis solaribus** ...(Augusta, 1612).

Mark Welser duumvir of Ausburg and friend of Galileo, the sixth of January sends to Galileo the booklet accompanied by a letter where he says, inter alia, «... Now others follow your lead with the greater courage, knowing that once you have broken the ice for them it would indeed be base not to press so happy and honorable an undertaking. See, then, what has arrived from a friend of mine; and if it does not come to you as anything really new, as I suppose, nevertheless I hope you will be pleased to see that on this side of the mountains also men are not lacking who travel in your footsteps...».[17]

---

[15]Ibidem (928–935), (Drake, p. 172).

[16]**Istoria e Dimostrazioni intorno alle macchie solari e loro accidenti** ... dal Sig. Galileo Galilei linceo—In Roma, MDCXIII.

[17]E. N. XI, p. 257 (Drake, p. 89).

Welser hints at the fact that Galileo previously (certainly, the last year) had dealt with the sunspots and then urges him to reply.

In the second edition of his essay **Intorno alle cose, che stanno in su l'acqua, e che in quella si muovono**[18] Galileo (in the current year) adds the phrase «Continued observations have finally assured me that such spots are materials contiguous to the sun's body, there continually produced in numbers and then dissolved, some in shorter and some in longer times, being carried around by rotation of the sun itself, which completes its period in about a lunar month a great event, and even greater for its consequences».[19]

The subsequent year he publishes, in the shape of three letters to the friend Welser, the work **Istoria e Dimostrazioni intorno alle macchie solari e loro accidenti** (which we have already quoted) in dispute with Scheiner. It is here that, insisting about the solar rotation, the following passage appears «For I seem to have observed that physical bodies have physical inclination to some motion (as heavy bodies downward), which motion is exercised by them through an intrinsic property and without need of a particular external mover, whenever they are not impeded by some obstacle. And to some other motion they have a repugnance (as the same heavy bodies to motion upward), and therefore they never move in that manner unless thrown violently by an external mover. Finally, to some movements they are indifferent, as are these same heavy bodies to horizontal motion, to which they have neither inclination (since it is not toward the center of the earth) nor repugnance (since it does not carry them away from that center).

And therefore, all external impediments removed, a heavy body on a spherical surface concentric with the earth will be indifferent to rest and to movements toward any part of the horizon. And it will maintain itself in that state in which it has once been placed; that is, if placed in a state of rest, it will conserve that; and if placed in movement toward the west (for example), it will maintain itself in that movement.

Thus a ship, for instance, having once received some impetus through the tranquil sea, would move continually around our globe without ever stopping; and placed at rest it would perpetually remain at rest, if in the first case all extrinsic impediments could be removed, and in the second case no external cause of motion were added».[20]

In this occasion too, Galileo seems to repeat the same enunciation of **Le Mechaniche** and of **De Motu**. But here there is a fundamental "addition". Galileo says that the moveable «will maintain itself in that state in which it has once been placed; that is, if placed in a state of rest, it will conserve that; and if placed in movement … it will maintain in that movement».

---

[18]Discorso al Serenissimo Don Cosimo II Gran Duca di Toscana **Intorno alle cose, che stanno in su l'acqua, e che in quella si muovono** di Galileo Galilei …. Seconda Editione, In Firenze, Apresso Cosimo Giunti. MDCXII.

[19]E. N. IV, p. 64 (Drake, p. 20).

[20]E. N. V, pp 134–135 (Drake, pp 113–114).

It is the first time that the statement appears that, if there are not external applied forces, the moveable will conserve his state of rest or of motion. Even if it is not explicitly said, it is obvious enough that in the case of motion, the moveable will continue to move with the velocity he had at the moment when he has been deprived by applied external forces. That is, the motion will be uniform.

On the fact that this uniform motion occurs on the surface of a sphere (having supposed the Earth of a spherical shape) and therefore it will be a circular motion, we shall come back in the next section by facing the problem in all its complexities.[21]

The origin of this "addition" to the "usual" enunciation must be imputed to the important and fundamental improvements in the study of motion, obtained in the Paduan years. These improvements, even if not published, were known to his disciples and his correspondents.

It is enough to think of the well-known correspondence with Paolo Sarpi of October 1604,[22] of the letter of Benedetto Castelli of April 1607[23] and, at last, of the letters of Galileo to Antonio De' Medici of February 1609[24] and to Belisario Vinta of May 1610.[25] In the latter, he proudly lists the works being completed regarding mechanics and from this one obviously deduces the conclusion that, not only he has not neglected the studies about motion in Padua, but also that he intends to continue them at Florence.

After the dispute on the sunspots, Galileo will have again the opportunity of competing in a dispute (with the Jesuit Orazio Grassi) on the appearance of three comets (sighted in November 1618). The conclusive work will be published in 1623 under the title **The Assayer**.

In this work there are no references to the inertia principle and thus, it lies outside our task. But we cannot get out of reminding that it is just in this work that appears the well-known passage «Philosophy is written in this grand book, the universe, which stands continually open to our gaze. But the book cannot be understood unless one first learns to comprehend the language and read the letters in which it is composed. It is written in the language of mathematics, and its characters are triangles, circles, and other geometric figures without which it is

---

[21]It must be absolutely pointed out that, when Galileo writes the enunciation on **Istoria e Dimostrazioni** ..., he is surely a convinced Copernican and, therefore, does not consider the Earth as the centre of the universe anymore. Therefore the circular motion of a moveable not subject to external forces around the Earth is no more the motion around the centre of the world but rather the motion of a moveable which has its own "gravity", even if one does not know how to justify this gravity yet.

[22]E. N. X, pp 114–116. This is the occasion in which Galileo unveils that he has obtained the result that «.... The spaces covered by the natural motion are in a double proportion with the times ...». We shall tell again of this in the next chapter.

[23]E. N. X, pp 169–171.

[24]E. N. X, pp 228–230 (It is the letter on the motion of the projectiles).

[25]E. N. X, pp 348–353, where he announces that among the works he has to accomplish («da condurre a fine») there are three books De Motu locali, a completely new science («scienza interamente nuova ...»).

humanly impossible to understand a single word of it; without these, one wanders about in a dark labyrinth».[26]

What above, one can say, synthesizes Galileo's scientific creed.

4.4   The Inertia Principle in the Dialogue and in the Discourses

The rule we have adopted for our work, that is of going backwards for seeking in Galileo's works those elements which later on converged in the classical mechanics, obliges us to extract from it those passages particularly meaningful for being interpreted as those bricks that will be used in the Newtonian building.

The task results particularly complicated in the case of the **Dialogue** and of the **Discourses**. The interval (6 years) elapsed between the publication of the two works does not allow to examine them in chronological sequence, at least for the inertia principle; in addition, considering the fact that, as Clavelin remarks, Galileo omits in the **Dialogue** some results already obtained, and in the meantime omits others in the **Discourses** (for instance the principle of relativity).

However, one must consider «…. the change in the conception of inertia along with the new way of regarding the effect of a constant force acting upon a particle of matter probably constitutes the most important element of all in the transition from ancient and medieval to classical science, which is our subject. Moreover, the law of inertia is not just a detail of the new world-picture, but one of the foundations underlying the most essential parts of the System.

That this change was very largely brought about by Galileo is beyond dispute; so is the fact that one cannot get an insight into its evolution in a more effective way than by studying his works. But then it must also be evident that the limitations, uncertainties, and inconsistencies to be traced in his reasonings will become of great significance as symptoms of the difficulties that had to be surmounted before a full understanding of inertia became possible».[27]

Let us discuss three passages, one from the **Dialogue** and two from the **Discourses**, considered by J.A. Coffa[28] particularly significant[29] for discussing of the "circular inertia", in opposition to those (A. Koyré and continuators) he calls "circularists".

---

[26]See: E. N. VI (Drake p. 237–238).

[27]E.J. Dijksterhuis—**The Mechanization of the World Picture**—Oxford at the Clarendon Press, 1961, pp 347–348.

[28]J.A. Coffa: **Galileo's concept of inertia**—Physis—X, 261–281, 1968.

[29]Unfortunately, the recourse to the practice of quoting passages considered "significant", even if inescapable, implies the risk of "partial" interpretations due (by means of the choice of the phrases quoted) to the removal of the context. In the case of the work of Galileo, and in particular for the **Dialogue**, this risk is particularly present since not always the "mechanical" enunciation is an end in itself, but is often in service of the construction and justification of a Copernican cosmology.

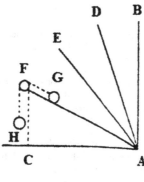

Fig. 4.2

Here are, in sequence, the three passages.

«We may therefore suppose it to be true that in the ordinary course of nature a body with all external and accidental impediments removed travels along an inclined plane with greater and greater slowness according as the inclination is less, until finally the slowness comes to be infinite when the inclination ends by coincidence with the horizontal plane.

We may likewise suppose that the degree of velocity acquired at a given point of the inclined plane is equal to the velocity of the body falling along the perpendicular to its point of intersection with a parallel to the horizon through the given point of the inclined plane.[30]

And if these two propositions be true, it follows necessarily that a falling body starting from rest passes through all the infinite gradations of slowness; and that consequently in order to acquire a determinate degree of velocity it must first move in a straight line, descending by a short or long distance according as the velocity to be acquired is to be lesser or greater, and according as the plane upon which it descends is slightly or greatly inclined. Hence a plane may be given so small an inclination that to acquire in it the assigned degree of velocity, a body must first move a very great distance and take a very long time.

In the horizontal plane no velocity whatever would ever be naturally acquired, since the body in this position will never move. But motion in a horizontal line which is tilted neither up nor down is circular motion about the center; therefore circular motion is never acquired naturally without straight motion to precede it; but, being once acquired, it will continue perpetually with uniform velocity.»[31]

«The better to explain this, let the line *AB* (Fig. 4.2) be assumed to be erected vertically on the horizontal *AC*, and then let it be tilted at different inclinations with respect to the horizontal, as at *AD, AE, AF*, etc. I say that the impetus of the heavy body for descending is maximal and total along the vertical *BA*, is less than that along *DA*, still less along *EA*, successively diminishes along the more inclined *FA*, and is finally completely extinguished on the horizontal *CA*, where the moveable is found to be indifferent to motion and to rest, and has in itself no inclination to move in any direction, nor yet any resistance to being moved.

Thus it is impossible that a heavy body (or combination thereof) should naturally move upward, departing from the common center toward which all heavy bodies mutually converge [*conspirano*]; and hence it is impossible that these be moved spontaneously except

---

[30]Here Galileo alludes his "postulate" which we shall deal with in the next chapter.
[31]E. N. VII, pp 52–53 (Drake, pp 31–32).

with that motion by which their own center of gravity approaches the said common center. Whence, on the horizontal, which here means a surface [everywhere] equidistant from the said [common] center, and therefore quite devoid of tilt, the impetus or momentum of the moveable will be null.»[32]

«SIMP. To these difficulties I add some more. One is that we assume the [initial] plane to be horizontal, which would be neither rising nor falling, and to be a straight line as if every part of such a line could be at the same distance from the center, which is not true. For as we move away from its midpoint towards its extremities, this [line] departs ever farther from the center [of the earth], and hence it is always rising.

One consequence of this is that it is impossible that the motion is perpetuated, or even remains equable through any distance; rather, it would be always growing weaker. Besides, in my opinion it is impossible to remove the impediment of the medium so that this will not destroy the equability of the transverse motion and the rule of acceleration for falling heavy things. All these difficulties make it highly improbable that anything demonstrated from such fickle assumptions can ever be verified in actual experiments.

SALV. All the difficulties and objections you advance are so well founded that I deem it impossible to remove them. For my part, I grant them all, as I believe our Author would also concede them. I admit that the conclusions demonstrated in the abstract are altered in the concrete, and are so falsified that horizontal [motion] is not equable; nor does natural acceleration occur [exactly] in the ratio assumed; nor is the line of the projectile parabolic, and so on. But on the other hand, I ask you not to reject in our Author what other very great men have assumed, despite its falsity.

The authority of Archimedes alone should. satisfy everyone; in his book **On Plane Equilibrium** [*Mecaniche*], as in the first book of his **Quadrature of the Parabola**, he takes it as a true principle that the arm of a balance or steelyard lies in a straight line equidistant at all points from the common center of heavy things, and that the cords to which [balance-]weights are attached hang parallel to one another.

These liberties are pardoned to him by some for the reason that in using our instruments, the distances we employ are so small in comparison with the great distance to the center of our terrestrial globe that we could treat 1 min of a degree at the equator as if it were a straight line, and two verticals hanging from its extremities as if they were parallel. Indeed, if such minutiae had to be taken into account in practical operations, we should have to commence by reprehending architects, who imagine that with plumb-lines they erect the highest towers in parallel lines.

Here I add that we may say that Archimedes and others imagined themselves, in their theorizing, to be situated at infinite distance from the center. In that case their said assumptions would not be false, and hence their conclusions were drawn with absolute proof. Then if we wish later to put to use, for a finite distance [from the center], these conclusions proved by supposing immense remoteness [therefrom], we must remove from the demonstrated truth whatever is significant in [the fact that] our distance from the center is not really infinite, though it is such that it can be called immense in comparison with the smallness of the devices employed by us ....»[33]

Coffa comments: «.... Galileo seems to reach an almost perfect formulation of rectilinear inertia, by stating that velocity is naturally conserved in a horizontal

---

[32]E. N. VIII, p. 215 (Drake, p. 172). This passage will be taken again in the Sect. 5.4.

[33]E. N. VIII, pp 274–275 (Drake, pp 223–224).

plane. Yet, rather surprisingly, he normally goes on to qualify these statements as only approximately valid, making it clear that whenever he mentions horizontal surfaces, he in fact means spheres concentric with the earth.»

By following the subsequent remarks of Coffa, let us analyze in the substance the proposed passages. First of all, one must state in advance that, differently of what was exposed in **De Motu** and in **Le Mechaniche**, now the motion along the inclined planes is correctly a uniformly accelerated motion. When the inclination decreases, the acceleration of a body which descends decreases and, alternatively, the deceleration of a body pushed upward decreases.

For continuity, a zero acceleration is due to the motion on a horizontal plane, therefore a constant velocity. But the horizontal plane, actually, changes from point to point, since the "horizontal" plane which counteracts at any point the gravity is that which is tangent to the spherical surface of the Earth.

Therefore, when one speaks of a horizontal plane, one has to do only with a first approximation. Coffa says that this is what, substantially, Galileo asserts: all this is rigorously consistent with the Newtonian mechanics, then the "circularists" cannot deny its evidence. The heavy body is subject (in modern language) to a central field of force, therefore in the "horizontal" plane (which counteracts the field) the moveable is at rest or moves with a motion which would be uniform rectilinear if the "horizontal" plane would remain always the same. Since, actually, the plane changes from point to point always remaining tangent to the (spherical) surface of the Earth, the moveable will go with constant velocity along a circumference having its centre in the centre of the Earth.

Therefore, Galileo is not stating a "circular" inertia principle, but is instead establishing what is the surface on which the linear velocity is conserved. It is quite equivalent that the gravity of the moveable is an intrinsic quality of the moveable or is due to an external action; the role of the "horizontal" plane is that of "counteracting" this force (weight or gravity, whatever it is). Rightly Coffa adds: «The first think to be said about this "mental experiment" is that the kind of "inertia" that we infer by this procedure is no inertia at all».[34]

The passage we have reported first (whose final part represents the third passage quoted by Coffa) belongs to the first day of the **Dialogue** and, in our opinion, it is not possible to give a correct interpretation of it if one does not consider it immersed in the context. In the first day of the **Dialogue**, Galileo establishes the preliminary remarks for being able to treat and justify in the second day the diurnal motion of the Earth. In fact, the first day begins with the exposition of Aristotle's ideas and then passes, on part of the alter ego Salviati, to expose the ideas of "our common friend Lincean Academician" (i.e. Galileo) on the accelerated motion which must (starting from the rest) pass «through all the infinite gradation of slowness without pausing in any of them».

We shall not dwell upon the accelerated motion, which will be treated in the next chapter, but we confirm that the aim of Galileo, in this case, as it seems to

---

[34]J.A. Coffa, loc. cit. p. 267.

us, is not that of enunciating a "circular" inertia principle but that of insisting that «circular motion can never be acquired naturally without straight motion preceding it». The aim of his discourse is that of arriving to say how the "divine Architect" has created all «those globes which we behold continually revolving» and has first impressed on them a rectilinear motion which then in various pre-established points for the various planets has been transformed in a circular motion around the centre (the Sun).

That is, what may appear a discourse purely regarding the mechanics of the "local motion" is really propaedeutic to the Galilean (Copernican) cosmology. Therefore, in our opinion, the discussion of a passage of the **Dialogue**[35] (still in the first Day) made by Koyré in his essay "**Galilée et la Loi d'Inertie**"[36] appears rather exploitable (i.e. in service of the thesis that the inertia principle must not be credited to Galileo but, in sequence, to Gassendi and Descartes).

In the Second passage (from the **Discourses**), where the usual reference to the model of the planes with different inclinations is repeated in short, it appears evident that «the horizontal ... here means a surface [everywhere] equidistant from the said [common] center, and therefore quite devoid of tilt ...».

The centre is the centre of the Earth: Galileo is not yet Newton and for him the gravity is not universal. The gravity is a property of the bodies which can be removed only by rescuing them from the tendency to move toward the centre of the Earth.

In the third passage (still from the **Discourses**) one wants to emphasize, by basing on an example of Archimedes and therefore on his authority, that in reality the spherical surface on which the gravity can be made vanish has a radius so great that even on great distances we can consider a body on it as it was on a plane. On this "plane", a body not acted upon by any external action remains at rest or moves with uniform motion. The concept is clearly expressed in this passage we quote from the **Dialogue**:

> «SALV. I do not want you to declare or reply anything that you do not know for certain. Now tell me: Suppose you have a plane surface as smooth as a mirror and made of some hard material like steel. This is not parallel to the horizon, but somewhat inclined, and upon it you have placed a ball which is perfectly spherical and of some hard and heavy material like bronze. What do you believe this will do when released? Do you not think, as I do, that it will remain still?
>
> SIMP. If that surface is tilted?
>
> SALV. Yes, that is what was assumed.
>
> SIMP. I do not believe that it would stay still at all; rather, I am sure that it would spontaneously roll down.
>
> SALV. Pay careful attention to what you are saying, Simplicio, for I am certain that it would stay wherever you placed it.

---

[35]E. N. VII, p. 43.

[36]A. Koyré: **Études Galiléennes**—III Galilée et la loi d'inertie—Hermann, 1939, pp 49–50.

SIMP. Well, Salviati, so long as you make use of assumptions of this sort I shall cease to be surprised that you deduce such false conclusions.

SALV. Then you are quite sure that it would spontaneously move downward?

SIMP. What doubt is there about this?

SALV. And you take this for granted not because I have taught it to you indeed, I have tried to persuade you to the contrary but all by yourself, by means of your own common sense.

SIMP. Oh, now I see your trick; you spoke as you did in order to get me out on a limb, as the common people say, and not because you really believed what you said.

SALV. That was it. Now how long would the ball continue to roll, and how fast? Remember that I said a perfectly round ball and a highly polished surface, in order to remove all external and accidental impediments. Similarly I want you to take away any impediment of the air caused by its resistance to separation, and all other accidental obstacles, if there are any.

SIMP. I completely understood you, and to your question I reply that the ball would continue to move indefinitely, as far as the slope of the surface extended, and with a continually accelerated motion. For such is the nature of heavy bodies, which *vires acquirunt eundo,* and the greater the slope, the greater would be the velocity.

SALV. But if one wanted the ball to move upward on this same surface, do you think it would go?

SIMP. Not spontaneously, no; but drawn or thrown forcibly, it would.

SALV. And if it were thrust along with some impetus impressed forcibly upon it, what would its motion be, and how great?

SIMP. The motion would constantly slow down and be retarded, being contrary to nature, and would be of longer or shorter duration according to the greater or lesser impulse and the lesser or greater slope upward.

SALV. Very well; up to this point you have explained to me the events of motion upon two different planes. On the downward inclined plane, the heavy moving body spontaneously descends and continually accelerates, and to keep it at rest requires the use of force. On the upward slope, force is needed to thrust it along or even to hold it still, and motion which is impressed upon it continually diminishes until it is entirely annihilated. You say also that a difference in the two instances arises from the greater or lesser upward or downward slope of the plane, so that from a greater slope downward there follows a greater speed, while on the contrary upon the upward slope a given movable body thrown with a given force moves farther according as the slope is less.

Now tell me what would happen to the same movable body placed upon a surface with no slope upward or downward.

SIMP. Here I must think a moment about my reply. There being no downward slope, there can be no natural tendency toward motion; and there being no upward slope, there can be no resistance to being moved, so there would be an indifference between the propensity and the resistance to motion. Therefore it seems to me that it ought naturally to remain stable. But I forgot; it was not so very long ago that Sagredo gave me to understand that that is what would happen.

SALV. I believe it would do so if one set the ball down firmly. But what would happen if it were given an impetus in any direction?

SIMP. It must follow that it would move in that direction.

SALV. But with what sort of movement? One continually accelerated, as on the downward plane, or increasingly retarded as on the upward one?

SIMP. I cannot see any cause for acceleration or deceleration, there being no slope upward or downward.

SALV. Exactly so. But if there is no cause for the ball's retardation, there ought to be still less for its coming to rest; so how far would you have the ball continue to move?

SIMP. As far as the extension of the surface continued without rising or falling.

SALV. Then if such a space were unbounded, the motion on it would likewise be boundless? That is, perpetual?

SIMP. It seems so to me, if the movable body were of durable material.

SALV. That is of course assumed, since we said that all external and accidental impediments were to be removed, and any fragility on the part of the moving body would in this case be one of the accidental impediments.

Now tell me, what do you consider to be, the cause of the ball moving spontaneously on the downward inclined plane, but only by force on the one tilted upward?

SIMP. That the tendency of heavy bodies is to move toward the center of the earth, and to move upward from its circumference only with force; now the downward surface is that which gets closer to the center, while the upward one gets farther away.

SALV. Then in order for a surface to be neither downward nor upward, all its parts must be equally distant from the center. Are there any such surfaces in the world?

SIMP. Plenty of them; such would be the surface of our terrestrial globe if it were smooth, and not rough and mountainous as it is. But there is that of the water, when it is placid and tranquil.

SALV. Then a ship, when it moves over a calm sea, is one of these movables which courses over a surface that is tilted neither up nor down, and if all external and accidental obstacles were removed, it would thus be disposed to move incessantly and uniformly from an impulse once received?»[37]

Another excerpt of the **Dialogue**, particularly emphasized by Stillman Drake[38] (always in debate with the "circularist" Koyré) and belonging to the discussion on the diurnal rotation of the Earth is the following:

«Salviati (addressed to Simplicio) … Up to this point you knew all by yourself that the circular motion of the projector impresses an impetus upon the projectile to move, when they separate, along the straight line tangent to the circle of motion at the point of separation, and that continuing with this motion, it travels ever farther from the thrower. And you have said that the projectile would continue to move along that line if it were not inclined downward by its own weight, from which fact the line of motion derives its curvature. It seems to me that you also knew by yourself that this bending always tends toward the center of the earth, for all heavy bodies tend that way….»[39]

---

[37]E. N. VII, pp 171–174 (Drake, pp 169–172).

[38]Stillman Drake: **The question of circular inertia**—Physis, X, 282–298, 1968.

[39]E. N. VII, p. 220 (Drake, p. 225).

We add at last, to proof that the gravity of the bodies is always considered as the obstacle to the perpetuity of the uniform rectilinear motion, the following passage from the incipit of the fourth day of the **Discourses** (devoted to the motion of the projectiles):

> «*I mentally conceive of some moveable projected on a horizontal plane, all impediments being put aside. Now it is evident from what has been said elsewhere at greater length that equable motion on this plane would be perpetual if the plane were of infinite extent, but if we assume it to be ended, and [situated] on high, the moveable (which I conceive of as being endowed with heaviness), driven to the end of this plane and going on further, adds on to its previous equable and indelible motion that downward tendency which it has from its own heaviness....*»[40]

This text (which we shall take again in due course, since the parabolic motion of the projectiles is theorized in its continuation) is the last published version (1638), with Galileo still alive, of the "restricted" inertia principle.

To us, this diction seems to be correct in the sense that Galileo says what "would happen" if it was possible to remove the gravity of the bodies. In the same time, we consider erroneous and misleading the diction "circular". Unfortunately, the charisma of Koyré has given birth to a crowd of epigones.

We shall see in the next section that Galileo was gone further (i.e. succeeding in inventing an experiment in which the gravity of the bodies is removed) but was not in time for delivering, to the publisher Lodowijk Elzevier (from Leiden), the original text of what will be called "sixth Day".

4.5 The Inertia Principle in the "Sixth Day"

The writing we want to deal with in this section is of fundamental importance for the subject we are treating, i.e. the attribution to Galileo of the formulation of the inertia principle (even if not in the definite form established by Newton). For this, it is convenient to investigate, as far as possible, the circumstances of its origin and conservation. In fact, we anticipate that it was published (printed) for the first time in 1718. Let us begin from the beginning, i.e. from what told by Viviani (1622–1703) in its **Life of Galileo** (1654):

> «During the time of thirty months in which I lived continually with him until the last days of his life, since he was very often tormented by shooting pains in his limb, which took him his sleep and rest, by a perpetual tingle at the eyelids, which was of unbearable vexation, and by the other affliction due to the ripe old age, grown weary by the so many studies and watches of the years before, he never could apply himself to handwrite the other works that were already thought and assimilated in his mind, but not yet written, as he wished to do. He had thought (since the Dialogues of the two New Sciences were already published) of forming two Days to be added to the other four; and in the first he intended to insert, besides the two aforesaid demonstrations, many new considerations and thoughts on various passages of the already printed Days, ... and finally in the last Day [he

---

[40]E. N. VIII, p. 268—This passage (italic) is in Latin in the original, under the usual artifice by means of which Salviati reports "the text of our Author".

intended] to promote another new science, dealing with geometric progress with the admirable force of percussion, which he said to have discovered himself and being able to demonstrate very keen and hidden conclusions, which greatly overcame all other his already published inquiries.»[41]

Twenty years later, Viviani comes back on the subject (in the "Ragguaglio delle ultime Opere di Galileo", from p. 85 to p. 106 of the **Quinto Libro degli elementi di Euclide**).

We extract from the narration of Viviani:

«In the hands of him[42] (who together with Torricelli, and with me had attended the illness, and the death of Galileo his Father, occurred the eight of January 1642) I saw, besides the Original proofs of the already printed Works, also those of several letters, and talks, written by Galileo in different times in occasions of answering or giving full details, or giving his opinion on questions put to him, or the like. He was pleased to give me a copy of them, also by dictating many of them to me when, and very frequently, I was at his home: even if, then, I saw most other copies of them, also handwritten, going around … The third writing he dictated to me, is another beginning of a new meeting, entitled Ultimate, and maybe so-called by Galileo before he conceived to reduce also the notes to his opponents into the form of a Dialogue. In this Meeting Galileo introduces (as usual) as conversation partners Salviati and Sagredo, by excluding Simplicio, and putting as the third that aforementioned Signor Paolo Aproino who was his Auditor of Mathematics in Padua. This opening is handwritten as a Dialogue on about six sheets, where some experiments which Galileo performed already in the time he was there as lecturer are explained, when he was investigating the measurement of the force of the percussion (that at last he considered as infinite); and after having explained the experiments, Galileo wanted to deal mathematically with the matter in all the remaining part of the meeting. He wanted to call it third Science, after the two he had already promoted, and to end with this the publication of the remainder of his most elaborated labours, as this would have happened, about which he himself said to have spent many thousands of hours speculating, and philosophizing, and finally having obtained further knowledge from our first conceptions, but new, and to be admired for its novelty.

At last, from what can be deduced from the aforementioned note of Galileo of January 1638, of all which remained to be written and published, in another Dialogue (which will have been the seventh, besides the first four of the chief systems) one had to include all those notes, remarks, and replies, called by him *postille*, regarding the most important passages written by those we who were against him …. But coming back to the copy I have, entitled Last Meeting, since in some passages I had some uncertainties, I was helped to check it with his own original by the Merciful Signor Cosimo, son of the aforesaid Signor Vincenzio, and noble grandson of Galileo, before he left Florence for passing in service of the Merciful Eminence Signor Cardinal Barbarigo, my most Benign, and most Revered Lord, and in that occasion he told me that he himself had already given off other copies.»[43]

It is just of the "third meeting", "other beginning of a new meeting, entitled Ultimate" that we have to deal with.

Galileo, in the past, had added (to the end of his work **Le Mechaniche**) a short writing entitled "Della forza della percossa"[44] and, as it seems, he dealt no more

---

[41]Vincenzio Viviani: **Vita di Galileo** a cura di Bruno Basile, op. cit. pp 68–69. (Our translation).

[42]Vincenzio Galilei, son of Galileo.

[43]Vincenzio Viviani: **Quinto Libro degli Elementi di Euclide**, ….op. cit. pp 102–104. (Our translation).

[44]See: **Le Mecaniche**, op. cit., pp 76–77.

with the subject. In the imminence of the publication of the **Discourses**, he had been solicited by various persons to go back to the question. He did this and dictated (as it seems testified) to the Florentine priest Marco Ambrogetti the writing that Viviani calls the Ultimate.

It is not clear why Viviani did not made include this Day on the force of percussion in the Bolognese edition (of the publisher Dozza) of 1655–56 and neither it is clear how the manuscript arrived in the hands of the Florentine publisher for the edition of 1718.[45] The fact remains that the handwritten copies of which Viviani speaks have been lost.

Further, the editor, being already the Day "On the definition of the Euclid's proportions" classified in the preceding Bolognese edition as "fifth" and being the new manuscript called Ultimate, called this "Sixth Day", whereas Galileo had considered it as the fifth and last of the **Discourses**.

The misadventures of this most important writing do not finish here. The various editions of the **Discourses** which followed one the other starting from the one of 1718, seldom comprised the sixth Day, only reporting, as if it was for a philological scruple, the content of the Elzevir edition of 1638. The title may have had influence in misleading the scholars, too. In fact, J. Coffa says: «…., a crucial passage in the Sixth Day of the **Discourses** whose relevance to this topic seems to have been, so far, overlooked».[46]

The only one after Coffa (at least to the best of our knowledge) who went on years later with the talk on the passage of the sixth Day and emphasized its importance has been Roberto Vergara Caffarelli.[47]

We report first the text, by postponing all considerations. Obviously, we limit the quotations to the bare minimum, being exclusively interested in the inertia principle. The reader interested in the other aspects may find an extensive treatment in the monograph of Vergara Caffarelli.

«… I want you to imagine a solid weight of, say, 1000 pounds, placed on a plane that sustains it (see Fig. 4.3). Next, I want you to think of a rope lied to this weight and led over a pulley fixed high above. Here it is evident that when force is applied by pulling down on the end of the rope, it will always meet with quite equal resistance in raising the weight; that is, the opposition of 1000 pounds of weight. For if from the end of the rope there were suspended another weight, equal to the first, equilibrium would be established; and being raised up without support from anything below, they would remain still; nor would this second weight descend and raise the first unless given some excess of weight. And if we rest the first weight on the said plane that sustains it, we can use other weights of varying heaviness (though each of them less than the weight sustained at rest) to test what the forces of different impacts are. [This is done] by tying such weights to the end of the rope

[45]**Opere di Galileo Galilei**, nobile fiorentino accademico linceo—Nuova Edizione con l'aggiunta di varij Trattati dell'istesso Autore e non più dati alle stampe. In Firenze MDCCXVIII. The sixth Day is in tome II, pp 693–710 (the edition is in three tomes).

[46]J. Coffa, loc. cit. p. 273.

[47]See: Roberto Vergara Caffarelli—**Il principio d'inerzia negli ultimi scritti di Galilei**— Cronos, 10, 63–88, 2007 and the beautiful book **Galileo Galilei and Motion—A Reconstruction of 50 Years of Experiments and Discoveries**—Springer and Società Italiana di Fisica, 2009. And the monograph **La Macchina di Galilei** (pp 46), s. d..

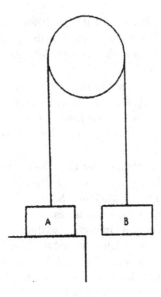

Fig. 4.3

and then letting them fall from a given height, observing what happens at the other end to that great solid that feels the pull of the falling weight, which pull will be to that large weight as a blow that would drive it upward.

Here, in the first place, it seems to me to follow that however small the falling weight, it should undoubtedly overcome the resistance of the heavy weight and lift it up. This consequence seems to me to be conclusively drawn from our certainty that a smaller weight will prevail over another, however much greater, whenever the speed of the lesser shall have, to the speed of the greater, a greater ratio than the weight of the greater has to the weight of the smaller; and this [always] happens in the present instance, since the speed of the falling weight infinitely surpasses the speed of the other, whose speed is nil when it is sustained at rest. But the heaviness of the falling solid is not nil in relation to that of the other, since we did not assume the latter to be infinite, or the former to be nil; hence the force of this percussent will overcome the resistance of that on which it makes its impact.»[48]

«Here, it seems to me, precisely the same thing takes place which happens to a heavy and perfectly round moveable placed on a very smooth plane, somewhat inclined; this will descend naturally by itself, acquiring ever greater speed. But if, on the other hand, anyone should wish to drive it upward from the lower part [of the plane], he would have to confer impetus on it, and this would be ever diminished and finally annihilated [in the rise].

If the plane were not inclined, but horizontal, then this round solid placed on it would do whatever we wish; that is, if we place it at rest, it will remain at rest, and given an impetus in any direction, it will move in that direction, maintaining always the same speed that it shall have received from our hand and having no action [by which] to increase or diminish this, there being neither rise nor drop in that plane.

And in this same way the two equal weights, hanging from the ends of the rope, will be at rest when placed in balance, and if impetus downward shall be given to one, it will always conserve

---

[48]E. N. VIII, pp 332–333 (Drake, pp 293–294).

this equably. Here it is to be noted that all these things would follow if there were removed all external and accidental impediments, as of roughness and heaviness of rope or pulleys, of friction in the turning of these about the axle, and whatever others there may be of these.»[49]

Before going deeply into the matter, we want to make some observations *a latere*.

First: to a physicist of the generation of the author, who has been familiar with the twentieth-century textbooks of "Experimental Physics", the description of the Galilean apparatus immediately recalls to mind the Atwood's machine[50] (1784). Effectively, one has to do with a pioneer version of a Atwood's machine.

Let us spent still a few words on this subject. The work of Atwood **A treatise of the rectilinear motion and rotation of bodies—with a description of original experiments** (Cambridge—1784), in section VII, devoted to the rectilinear motion of bodies subject to a constant force, quotes Galileo with regard to the theory of the motion along the inclined planes («the celebrated author of this theory»); but the treatise is essentially an exposition and an experimental verification of the Newtonian mechanics. Probably, Atwood knew the work of Galileo only indirectly and had not read the sixth Day (1718).

Second: Why, in the time in which the invention of anything was attributed to Galileo, no "patriot" has claimed the invention of that machine to Galileo? We have not found any trace of such a thing. Rightly, and correctly, Vergara Caffarelli says that Atwood has reinvented the machine of Galileo 159 years later.

Coming back to our principal point, it clearly emerges that Galileo has finally found the method for removing the obstacle of the gravity of the bodies, which now can move along the vertical with uniform rectilinear motion.

The inertia principle is no more "restricted", that is, limited to what would occur if the bodies had not the gravity. All this, inside a device invented for dealing with "the force of percussion", i.e. for establishing a theory of collisions which, as we now, will not be completed. In fact, Galileo will make a mistake in evaluating the velocity of falling bodies, since (we would say nowadays) he did not know the law of conservation of momentum.

At this point, one could think that the question is closed and then put, as one uses to say, a tombstone on the dispute of the "circularists". But the old Galileo can still fire off a shot and he does it with his usual weapon: the inclined plane. Let us see in fact, as he brings to a close his discourse:

> «It is further necessary to remember some true conclusions of which we spoke in past days in the treatise on motion. The first of these is that heavy bodies, in falling from a high point to a horizontal plane beneath, acquire equal degrees of speed whether their descent is made vertically or upon any of diversely inclined planes.

> For example, *AB* being a horizontal plane upon which the vertical *CB* is dropped from point *C* (Fig. 4.4), and other planes, diversely inclined, *C A, CD,* and *C E,* fall from the same C, we must understand that the degrees of speed of bodies falling from the high point C along any of the lines going from C to end at the horizontal are all equal. In

---

[49]E. N. VIII, pp 336–337 (Drake, pp 296–297).

[50]See, as an examle, O. D. Chwolson: **Traité de Physique—1: Mecanique ...** Paris, Hermann, 1908, pp 96–98.

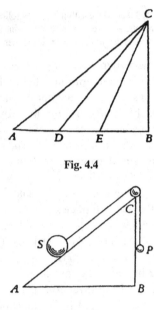

Fig. 4.4

Fig. 4.5

the second place, it is assumed that the impetus acquired at *A* by the body falling from the point *C* is such that it is exactly needed to drive the same falling body (or another one equal to it) up to the same height, from which we may understand that such force is required to raise that same heavy body from the horizontal to height *C*, whether it is driven from point *A, D, E,* or *B*.

Let us recall in the third place that the times of descent along the designated planes have the same ratio as the lengths of these planes, so that if the plane *AC*, for example, were double the length of *CE* and quadruple that of *CB,* the time of descent along *CA* would be double the time of descent along *CE* and four times that along *CB*. Further, let us recall that in order to pull the same weight over diverse inclined planes, lesser force will always suffice to move it over one which is more inclined [to the vertical] than over one less inclined, according as the length of the latter is less than the length of the former.

Now, these truths being supposed, let us take the plane *AC* to be, say, ten times as long as the vertical *CB*, and let there be placed on *AC* a solid, *S*, weighing 100 pounds (Fig. 4.5). It is manifest that if a cord is attached to this solid, riding over a pulley placed above the point *C*, and to the other end of this cord a weight of ten pounds is attached, which shall be the weight of *P*, then that weight *P* will descend with any small addition of force, drawing the weight S along the plane *AC*. Here one must note that the space through which the greater weight moves over the plane beneath it is equal to the space through which the small descending weight is moved; from this, someone might question the general truth applying to all mechanical propositions, which is that a small force does not overcome and move a great resistance unless the motion of the former exceeds the motion of the latter in inverse ratio of their weights.

But in the present instance the descent of the small weight, which is vertical, must be compared [only] with the vertical rise of the great solid S observing how much this is lifted vertically from the horizontal; that is, one must consider how much S rises in the vertical *BC*.»[51]

---

[51]E. N. VIII, pp 338–339 (Drake, pp 298–300).

The result obtained is: we now can think of a body released from any force, both "internal" (gravity) and external, and free to move in any direction (no more only vertical) with uniform rectilinear motion or to remain still.

The inertial principle can be enunciated, even if Galileo does not, also in the Newtonian form. Finally, we owe Galileo another recognition. All that we have told until here has been obtained on the basis of experiments really performed (in the Paduan period). Besides the account of Viviani, which sometimes may even be considered scantily objective, there are several testimonies on the skillfulness of Galileo in the construction of the experimental apparatus.

His description refers to apparatus effectively handled by him.

It is obvious that, in this last case, he relates by heart things of many years before. Anyhow, on Galileo "experimenter" and not "Platonist", as Koyré wanted him to be, we shall come back in the next chapter.

Before closing, it remains, as our duty, to give account of the "competition". Koyré supplies us with inertia principles, to be opposed to the Galileo's results: the enunciations of Gassendi and Descartes.

Let us begin from the first. Gassendi (1592–1655), who has studied thoroughly Galileo, but also Kepler and Gilbert, arrives to attribute the gravity to a kind of "attraction" by the Earth and gives a formulation of the principle of this type:

«Argumentum vero desumo ex aequabilitate illa motus horizontalis iam exposita; cum ille videatur aliunde non definere, nisi ex admissione motus perpendicularis; adeo ut, quia in illis spatiis nulla esset perpendicularis admistio, in quamcumque partem foret motus inceptus, horizontalis instar esset, et neque acceleraretur, retardareturve, neque proinde unquam desineret.»[52]

As to Descartes, usually one refers to the two works: : **Le Monde ou traité de la lumière** (published posthumous but written around the thirties) and **Principia Philosophiae** (Amsterdam, 1644). We remind that Descartes, born in 1596, died in 1650. We give here the text of the **Principia**:

«XXXVII—Prima lex naturae: quod unaquaeque res, quantum in se est, semper in eodem statu perseveret, sicque quod semel movetur, semper moveri pergat.»[53]

As one can see, here we are in presence of a "philosophical" enunciation (there is no indication of physical magnitudes). On the other hand, if one looks at the context in which this law is inserted, one meets with passages of the type «… ex

---

[52] «I adduce the proof from the uniformity of the horizontal motion already exposed; since this does not seem otherwise to finish, if the perpendicular motion is not added; from this, it follows that, since in those (empty) spaces there will not be addition of perpendicular motion, wherever the motion begins, it will be horizontal, neither will accelerate, nor will slow down, and then will never finish». From Petri Gassendi: **De Motu impresso a Motore translato**—Parisiis, MDCXLII. Epist. I, XVI, pp 62–63. (Our translation).

[53] «XXXVII—First law of nature: anything, for what is in it, always persists in the same state; and so, once it is in motion, continues always to move». See: Renati Des Cartes—**Principia Philosophiae**—Amstelodami, MDCXLIV, p. 54 or the modern edition **Ouvres de Descartes** (edited by Ch. Adam & P. Tannery)—VIII—Vrin, Paris—p. 62. (Our translation).

hoc solo, quod Deus diversimode moveris partes materiae, cum primum illas creavit, ...»[54] etc.

One has to do with a physics different from that of Galileo. The enunciation of Huygens (datable between 1669 and 1673) will be much less "philosophical", indeed almost Newtonian:

«I—Si gravitas non esset, neque aër motui corporum officeret, unum quodque eorum, acceptum semel motum continuaturum velocitate aequabili, secundum lineam rectam.»[55]

---

[54]René Descartes, op. cit. ibidem.

[55]Cristiani Hugenii **Horologium Oscillatorium sive de motu pendulorum ad horologia aptato—demonstrationes geometricae**—Parisiis, MDCLXXIII, p. 21. «*If there were no gravity, and if the air did not impede the motion of bodies, then any body will continue its given motion with uniform velocity in a straight line*». From: Cristian Huygens—**The Pendulum Clock or Geometrical Demonstrations Concerning the Motion of Pendula as applied to clocks**—Translated by R.J. Blackwell—The Iowa State University Press, 1986, p. 33.

Chapter 5
The Motion of Heavy Bodies and the Trajectory of Projectiles

The study of law of fall of heavy bodies and the consequent treatment of the uniformly accelerated motion are, as we know, the fundamental part of Galileo's mechanics. He has worked on it, so to speak uninterruptedly, during the Paduan period and after his return to Florence.

Yet, for reasons which can seem incomprehensible to us, may be because ever pressed by the circumstances of his Copernican battle, he has waited till the end, literally, assembling in a systematic and—in a way—rigorous form the results obtained, and giving them the form of a treatise.

Before of the **Discourses** (1638), in addition to the various communications and discussions which we find scattered in his plentiful correspondence, he anticipated, so to say, many of the conclusions already achieved in the Days of the **Dialogue**, drawn up in various intervals between 1624 and 1630. This in the second of the four Days, which is devoted to demonstrating that the traditional arguments against the Earth's diurnal motion are unfounded.

But it is in the third Day of the **Discourses** that we find the complete treatment of the law of fall and of the uniformly accelerated motion.

5.1 An Anticipation

In the second Day of the **Dialogue**, while discussing about the time that a "cannon ball" would take, falling from the moon's orbit to the Earth's centre, Galileo (Salviati) gives a first outline of his ideas on the motion of bodies:

«SALV. First of all, it is necessary to reflect that the movement of descending bodies is not uniform, but that starting from rest they are continually accelerated. This fact is known and observed by all, except the modern author mentioned,[1] who, saying nothing about

---

[1]Christopher Scheiner.

© Springer International Publishing Switzerland 2016
D. Boccaletti, *Galileo and the Equations of Motion*,
DOI 10.1007/978-3-319-20134-4_5

acceleration, makes the motion uniform. But this general knowledge is of no value unless one knows the ratio according to which the increase in speed takes place, something which has been unknown to all philosophers down to our time. It was first discovered by our friend the Academician, who, in some of his yet unpublished writings, shown in confidence to me and to some other friends of his, proves the following.

The acceleration of straight motion in heavy bodies proceeds according to the odd numbers beginning from one. That is, marking off whatever equal times you wish, and as many of them, then if the moving body leaving a state of rest shall have passed during the first time such a space as, say, an ell, then in the second time it will go three ells; in the third, five; in the fourth, seven; and it will continue thus according to the successive odd numbers. In sum, this is the same as to say that the spaces passed over by the body starting from rest have to each other the ratios of the squares of the times in which such spaces were traversed. Or we may say that the spaces passed over are to each other as the squares of the times.

SAGR. This is a remarkable thing that I hear you saying. Is there a mathematical proof of this statement?

SALV. Most purely mathematical, and not only of this, but of many other beautiful properties belonging to natural motions and to projectiles also, all of which have been discovered and proved by our friend. I have seen and studied them all, to my very great delight and amazement, seeing a whole new science arise around a subject on which hundreds of volumes have been written; yet not a single one of the infinite admirable conclusions within this science had been observed and understood by anyone before our friend.[2]

SAGR. You are taking away from me my desire to proceed with the discussions we have commenced, in order just to hear some of the demonstrations you hint of. So tell them to me at once, or at least give me your word that you will hold a special session with me, Simplicio being present if he should wish to learn the properties and attributes of the most basic effect in nature.

SIMP. Indeed I should; though as to what belongs to physical science, I do not believe it necessary to get down to minute details. A general knowledge of the definition of motion and of the distinction between natural and constrained motion, uniform and accelerated motion, and the like, is sufficient. For if these were not enough, I do not believe that Aristotle would have neglected to teach us everything that was lacking.

SALV. That might be. But let us waste no more time on this, for I promise to spend half a day on it separately for your satisfaction. Indeed, I now remember having once before promised you this same satisfaction. Getting back to our calculation, already begun, of the time in which a heavy body would fall from the moon's orbit all the way to the center of the earth, and in order not to proceed arbitrarily or at random, but with a rigorous method, let us first seek to make sure by experiments repeated many times how much time is taken by a ball of iron, say, to fall to earth from a height of one hundred yards.

SAGR. And taking for this purpose a ball of determinate weight, the same as that for which we shall make the computation of the time of descent from the moon.

SALV. That makes no difference at all, for a ball of one, ten, a hundred, or a thousand pounds will all cover the same hundred yards in the same time.

---

[2]Here, as one can see, Galileo does not escape his common attitude of selfpromotion (delegated to his alter ego Salviati) which sometimes degenerated into a denigration of his opponents. It is this habit that caused him the dislike of some historians (among the nearest to us, for instance A. Koestler in The **Sleepwalkers** and P. Feyerabend in **Against Method**).

SIMP. Oh, that I do not believe, nor does Aristotle believe it either; for he writes that the speeds of falling heavy bodies have among themselves the same proportions as their weights.

SALV. Since you want to admit this, Simplicio, you must also believe that a hundred-pound ball and a one-pound ball of the same material being dropped at the same moment from a height of one hundred yards, the larger will reach the ground before the smaller has fallen a single yard. Now try, if you can, to picture in your mind the large ball striking the ground while the small one is less than a yard from the top of the tower.

SAGR. I have no doubt in the world that this proposition is utterly false, but I am not quite convinced that yours is completely true; nevertheless I believe it because you affirm it so positively, which I am sure you would not do unless you had definite experiments or rigid proofs.

SALV. I have both, and when we deal separately with the subject of motion I shall communicate them to you. Meanwhile, in order not to break the thread again, let us suppose we want to make the computations for an iron ball of one hundred pounds which in repeated experiments falls from a height of one hundred yards in five seconds.»[3]

As to the fall of heavy bodies, Galileo limits himself to enunciate the law which connects the space with the square of the time, whereas the relation which links the velocity and the time in the uniformly accelerated motion is missing (we shall discuss it later on).

We omit the following calculation, in which Salviati dwells on numerical computations, however based on uncorrected data and on the hypothesis (obviously wrong) that the acceleration due to gravity is a constant (rather than following the inverse-square law as Newton will establish). On the contrary, the further passage is interesting, in which uses the example of the pendulum oscillations to state that the motion of a "cannon ball" within a canal going from one Earth pole to the other is undoubtedly an oscillatory motion.

We know that, if we study the motion of a mass point which moves diametrically within an ideal spherical Earth with constant density (like a hydrogen atom in the Thomson model) using the modern mechanics, we obtain that its motion is really a harmonic motion.

But, by inserting the current figures for the Earth's mass and radius and the universal gravitation constant, we obtain, obviously, a completely different result from that calculated by Galileo (Salviati) supposing a constant acceleration.[4]

«SALV. ...... Meanwhile I wish to set forth some conjectures, not to teach you anything new, but to take away from you a certain contrary belief and to show you how matters may stand. Have you not observed that a ball of lead suspended from the ceiling by a long, thin thread, when we remove it from the perpendicular and release it, will spontaneously pass beyond the perpendicular almost the same amount?

---

[3]E. N. VII, pp. 248–249; Drake, pp. 257–259.

[4]If we call r the distance of the mass point from the Earth's centre, in the above hypotheses the motion equation is $\frac{d^2r}{dt^2} + \frac{GM}{R^3} r = 0$ and, as a consequence, the oscillation period is $T = 2\pi \sqrt{\frac{R^3}{GM}}$, where M and R are the mass and the radius of Earth and G is the universal gravitation constant. It comes out $T \cong 5 \times 10^3$ s.

SAGR. I have indeed observed that, and I have seen (especially when the ball is very heavy) that it rises so little less than it descends that I have sometimes thought the ascending arc would be equal to the descending one, and wondered whether the oscillations could perpetuate themselves. And I believe that they would, if the impediment of the air could be removed, which, with its resistance to being parted, holds back a little and would impede the motion of the pendulum. But the hindrance is small indeed, as argued by the large number of vibrations made before the moving ball is completely stopped.

SALV. The motion would not perpetuate itself, Sagredo, even if the impediment of the air were completely removed, for there is another one which is much more recondite.

SAGR. And what is that? None other occurs to me.

SALV. It will please you very much to learn of it, but I shall tell it to you later; meanwhile, let us continue. I have put forth the observation of the pendulum so that you would understand that the impetus acquired in the descending arc, in which the motion is natural, is able by itself to drive the same ball upward by a forced *(violento)* motion through as much space in the ascending arc; by itself, that is, if all external impediments are removed.

I believe also that you understand without any trouble that just as in the descending arc the velocity goes on increasing to the lowest point of the perpendicular, so in the ascending arc it keeps diminishing all the way to the highest point. The latter speed diminishes in the same ratio in which the former is augmented, so that the degrees of speed at points equally distant from the lowest point are equal to each other.

From this it seems possible to me (arguing with a certain latitude) to believe that if the terrestrial globe were perforated through the centre, a cannon ball descending through the hole would have acquired at the centre such an impetus from its speed that it would pass beyond the centre and be driven upward through as much space as it had fallen, its velocity beyond the centre always diminishing with losses equal to the increments acquired in the descent»;[5]

After a detailed description of the motion of the "cannon ball" by Salviati and a successive confirmation by Sagredo of having understood everything, Salviati starts again:

SALV. «In accordance with your very swift and subtle comprehension, you have expressed the whole thing much more clearly than I did, and you also made me think of something else to add. For the increases in the accelerated motion being continuous, one cannot divide the ever-increasing degrees of speed into any determinate number; changing from moment to moment, they are always infinite. Hence we may better exemplify our meaning by imagining a triangle, which shall be this one, ABC (Fig. 5.1). Taking in the side AC any number of equal parts AD, DE, EF, and FG, and drawing through the points D, E, F, and G straight lines parallel to the base BC, I want you to imagine the sections marked along the side AC to be equal times. Then the parallels drawn through the points D, E, F, and G are to represent the degrees of speed, accelerated and increasing equally in equal times.

Now A represents the state of rest from which the moving body, departing, has acquired in the time AD the velocity DH, and in the next period the speed will have increased from the degree DH to the degree EI, and will progressively become greater in the succeeding times, according to the growth of the lines FK, GL, etc.

[5]E. N. VII, pp. 252–253; Drake, pp. 262–264.

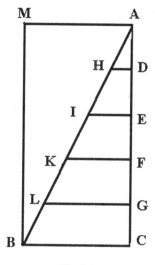

Fig. 5.1

But since the acceleration is made continuously from moment to moment, and not discretely *(intercisamente)* from one time to another, and the point A is assumed as the instant of minimum speed (that is, the state of rest and the first instant of the subsequent time AD), it is obvious that before the degree of speed DH was acquired in the time AD, infinite others of lesser and lesser degree have been passed through. These were achieved during the infinite instants that there are in the time DA corresponding to the infinite points on the line DA.

Therefore to represent the infinite degrees of speed which come before the degree DH, there must be understood to be infinite lines, always shorter and shorter, drawn through the infinity of points of the line DA, parallel to DH.

This infinity of lines is ultimately represented here by the surface of the triangle AHD. Thus we may understand that whatever space is traversed by the moving body with a motion which begins from rest and continues uniformly accelerating, it has consumed and made use of infinite degrees of increasing speed corresponding to the infinite lines which, starting from the point A, are understood as drawn parallel to the line HD and to IE, KF, LG, and BC, the motion being continued as long as you please.

Now let us complete the parallelogram AMBC and extend to its side BM not only the parallels marked in the triangle, but the infinity of those which are assumed to be produced from all the points on the side AC. And just as BC was the maximum of all the infinitude in the triangle, representing the highest degree of speed acquired by the moving body in its accelerated motion, while the whole surface of the triangle was the sum total of all the speeds with which such a distance was traversed in the time AG, so the parallelogram becomes the total and aggregate of just as many degrees of speed but with each one of them equal to the maximum BC.

This total of speeds is double that of the total of the increasing speeds in the triangle, just as the parallelogram is double the triangle. And therefore if the falling body makes use of the accelerated degrees of speed conforming to the triangle ABC and has passed over a certain space in a certain time, it is indeed reasonable and probable that by making use of the uniform velocities corresponding to the parallelogram it would pass with uniform motion during the same time through double the space which it passed with the accelerated motion.

SAGR. I am entirely persuaded. But if you call this a probable argument, what sort of thing would rigorous proofs be? I wish to Heaven that in the whole of ordinary philosophy there could be found even one proof this conclusive!.»[6]

The addition by Salviati wants to point out the fact that the velocity of the accelerated motion varies continuously and not discretely. This argument appears obvious and pleonastic to us but this was not the case at the time of Galileo when it was usual to proceed always by combining uniform motions (as the mentioned "modern author" Christopher Scheiner did). We shall discuss in 5.3 the argument that Salviati anticipates at the end of his speech and the relation with the Mertonian rule.

5.2 The Motion of Falling Bodies in the "Third Day" of the Discourses

After what we have called an "anticipation", let us see now how, in the final work, the matter is arranged in the form of a "treatise", equipped with "demonstrations".

The third Day of the **Discourses** is dedicated to the "local motion", both "natural" and "violent" (in accordance with the traditional subdivision of Aristotelian origin).

Galileo begins by announcing with pride the results he has achieved on problems which, nowadays, may appear simple and elementary, but at that time were substantially unsolved. Better to say that the existing solutions were conflicting with the common sense. One can say that until then no one had carried out experiments by performing measurements, neither on the (rectilinear) fall of bodies nor on the motion of projectiles.

Actually, on those problems the dust of more than fifteen centuries was lying. This is how Galileo summarizes his results:

«*ON LOCAL MOTION*[7]

We bring forward [promovemus] a brand new science concerning a very old subject.

There is perhaps nothing in nature older than MOTION, about which volumes neither few nor small have been written by philosophers; yet I find many essentials [symptomata] of it that are worth knowing which have not even been remarked, let alone demonstrated.

Certain commonplaces have been noted, as for example that in natural motion, heavy falling things continually accelerate; but the proportion according to which this acceleration takes place has not yet been set forth. Indeed no one, so far as I know, has demonstrated that the spaces run through in equal times by a moveable descending from rest maintain among themselves the same rule [rationem] as do the odd numbers following upon unity.

---

[6]E. N. VII, p. 256; Drake, pp. 265–267.

[7]For the parts of the text of the **Discourses** in Latin in the original, we shall make use of the italic type.

It has been observed that missiles or projectiles trace out a line somehow curved, but no one has brought out that this is a parabola.[8]

That it is, and other things neither few nor less worthy [than this] of being known, will be demonstrated by me, and (what is in my opinion more worthwhile) there will be opened a gateway and a road to a large and excellent science of which these labors of ours shall be the elements, [a science] into which minds more piercing than mine shall penetrate to recesses still deeper.

We shall divide this treatise into three parts. In the first part we consider that which relates to equable or uniform motion; in the second, we write of motion naturally acceler- ated; and in the third, of violent motion, or of projectiles.»[9]

After having exposed his treatment of the equable motion (i.e. uniform rectilin- ear, in the present definition), Galileo goes on to deal with the naturally acceler- ated motion (i.e. uniformly accelerated, in the present definition):

«*ON NATURALLY ACCELERATED MOTION*

Those things that happen which relate to equable motion have been considered in the pre- ceding book; next, accelerated motion is to be treated of.

And first, it is appropriate to seek out and clarify the definition that best agrees with that [accelerated motion] which nature employs. Not that there is anything wrong with invent- ing at pleasure some kind of motion and theorizing about its consequent properties, in the way that some men have derived spiral and conchoidal lines from certain motions, though nature makes no use of these [paths]; and by pretending these, men have laudably demon- strated their essentials from assumptions [ex suppositione].

But since nature does employ a certain kind of acceleration for descending heavy things, we decided to look into their properties so that we might be sure that the definition of accelerated motion which we are about to adduce agrees with the essence of naturally accelerated motion. And at length, after continual agitation of mind, we are confident that this has been found, chiefly for the very powerful reason that the essentials successively demonstrated by us correspond to, and are seen to be in agreement with, that which physi- cal experiments [naturalia experimenta] show forth to the senses.

Further, it is as though we have been led by the hand to the investigation of naturally accelerated motion by consideration of the custom and procedure of nature herself in all her other works, in the performance of which she habitually employs the first, simplest, and easiest means. And indeed, no one of judgment believes that swimming or flying can be accomplished in a simpler or easier way than that which fish and birds employ by nat- ural instinct.

Thus when I consider that a stone, falling from rest at some height, successively acquires new increments of speed, why should I not believe that those additions are made by the simplest and most evident rule? For if we look into this attentively, we can discover no simpler addition and increase than that which is added on always in the same way.

We easily understand that the closest affinity holds between time and motion, and thus equable and uniform motion is defined through uniformities of times and spaces; and

---

[8]It is worthwhile to remember that really, until then, Galileo had never published his discovery. At this regard, we also remember an "unpleasant incident" (see Appendix).

[9]E. N. VIII, p. 190; Drake, pp. 147–148.

indeed, we call movement equable when in equal times equal spaces are traversed. And by this same equality of parts of time, we can perceive the increase of swiftness to be made simply, conceiving mentally that this motion is uniformly and continually accelerated in the same way whenever, in any equal times, equal additions of swiftness are added on.

Thus, taking any equal particles of time whatever, from the first instant in which the moveable departs from rest and descent is begun, the degree of swiftness acquired in the first and second little parts of time [together] is double the degree that the moveable acquired in the first little part [of time]; and the degree that it gets in three little parts of time is triple; and in four, quadruple that same degree [acquired] in the first particle of time.

So, far clearer understanding, if the moveable were to continue its motion at the degree of momentum of speed acquired in the first little part of time, and were to extend its motion successively and equably with that degree, this movement would be twice as slow as [that] at the degree of speed obtained in two little parts of time. And thus it is seen that we shall not depart far from the correct rule if we assume that intensification of speed is made according to the extension of time; from which the definition of the motion of which we are going to treat may be put thus:

[DEFINITION]

I say that that motion is equably or uniformly accelerated which, abandoning rest, adds on to itself equal momenta of swiftness in equal times.»[10]

We must remark that, in the foregoing, Galileo makes use of both the term "gradus" (degree) and "momentum"[11] relating to velocity to indicate what for us is the instantaneous velocity. Since he had not to hand a "true" definition of velocity (for the reasons we have already recalled), obviously this allowed to make use of different terms to indicate the same physical magnitude. Moreover, it is to be remarked the permanence of a certain terminology going back to XIV century, for instance "intension velocitatis" and "extension velocitatis" (in the Latin text); and (still in the Latin text) the velocity is called alternatively "velocitas" or "celeritas".

Finally, it is particularly important the attention drawn by Galileo to the connection of the motion with the flow of time, till to anticipate the relation of proportionality between velocity and time in the uniformly accelerated motion which will be enunciated later.

5.2.1 Appendix—An "Unpleasant Incident"

The 31st of August 1632, Galileo, already harassed by the polemics following the publication of the **Dialogue**, received a letter from Bonaventura Cavalieri (his

---

[10]E. N. VIII, pp. 197–198; Drake, pp. 153–154.

[11]The term momentum (momento), in the course of time, in all Galilean work, unfortunately has been used in different occasions with different meanings. At this regard, as we have already recalled, a fundamental work is the study of Paolo Galluzzi; **Momento—Studi Galileiani—** Edizioni dell'Ateneo & Bizzarri Roma, 1979, which deals with the use of this term not only in the work of Galileo.

disciple and father of the order of Jesuati); there the disciple, giving notice of the publication of his last book **Lo Specchio Ustorio**,[12] inter alia said:

> «I have briefly touched the motion of projected bodies by showing that if the resistance of the air is excluded it must take place along a parabola, provided that your principle of the motion of heavy bodies is assumed that their acceleration corresponds to the increase of the odd numbers as they follow each other from one onwards. I declare, however, that I have learned in great parts from you what I touch upon in this matter, at the same time advancing myself a derivation of that principle.»[13]

The 11th of September Galileo, shocked by this announcement, writes to his friend Cesare Marsili in Bologna:

> «I have letters from Father Fra Buonaventura with the news that he had recently given to print a treatise on the burning mirror in which, as he says, he has introduced on an appropriate occasion the theorem and the proof concerning the trajectory of projected bodies in which he explains that it is a parabolic curve.
>
> I cannot hide from you, my dear Sir, that this news was anything but pleasant to me because I see how the first fruits of a study of mine of more than forty years, imparted largely in confidence to the said Father, should now be wrenched from me and how the flower shall now be broken from the glory which I hoped to gain from such long-lasting efforts, since truly what first moved me to speculate about motion was my intention of finding this path, which, although once found it is not very hard to demonstrate, still I, who discovered it, know how much labour I spent in finding that conclusion.»[14]

The 21st of September, Cavalieri writes to Galileo a letter in which he is contrite and justifies his behaviour with the following words:

> «I add that I truly thought that you had already somewhere written about it, as I have not been in the lucky situation to have seen all your works, and it has encouraged my belief that I realized how much and how long this doctrine has been circulated already, because Oddi has told me already ten years ago that you have performed experiments about that matter together with Sig.r Guidobaldo del Monte, and that also has made me imprudent so that I have not written you earlier about it, since I believed, in fact, that you do in no way bother about it but would rather be content that one of your disciples would show himself on such a favorable occasion as an adept of your doctrine of which he confesses to have learned it from you.»[15]

Galileo accepted the act of submission and **Lo Specchio Ustorio** was not burned (forgive the play on words).

---

[12]The work is: Bonaventura Cavalieri—**Lo Specchio Ustorio ovvero Trattato delle Settioni coniche, et di alcuni loro mirabili effetti**—in Bologna 1632. See also the recent edition edited by Enrico Giusti, Giunti, 2001. To the best of our knowledge an English translation (possible title **The Burning Mirror**) does not exist yet.

[13]E. N. XIV, p. 378 (15–23) (Translation by Renn et al. p. 56—the complete reference of the relevant paper will be given in Footnote 55).

[14]E. N. XIV, p. 386 (2–24) (Renn et al. pp. 56–57).

[15]E. N. XIV, pp. 394–395 (2–48) (Renn et al. p. 57).

5.3 Velocity and Space in the Uniformly Accelerated Motion

After some words of the dialogue, finally the alter ego Salviati enunciates the relation of proportionality between velocity and time in the uniformly accelerated motion:

«SALV. The present does not seem to me to be an opportune time to enter into the investigation of the cause of the acceleration of natural motion, concerning which various philosophers have produced various opinions, some of them reducing this to approach to the center; others to the presence of successively less parts of the medium [remaining] to be divided; and others to a certain extrusion by the surrounding medium which, in rejoining itself behind the moveable, goes pressing and continually pushing it out. Such fantasies, and others like them, would have to be examined and resolved, with little gain.

For the present, it suffices our Author that we understand him to want us to investigate and demonstrate some attributes [passiones] of a motion so accelerated (whatever be the cause of its acceleration) that the momenta of its speed go increasing, after its departure from rest, in that simple ratio with which the continuation of time increases, which is the same as to say that in equal times, equal additions of speed are made.

And if it shall be found that the events that then shall have been demonstrated are verified in the motion of naturally falling and accelerated heavy bodies, we may deem that the definition assumed includes that motion of heavy things, and that it is true that their acceleration goes increasing as the time and the duration of motion increases.

SAGR. By what I now picture to myself in my mind, it appears to me that this could perhaps be defined with greater clarity, without varying the concept, [as follows]: Uniformly accelerated motion is that in which the speed goes increasing according to the increase of space traversed.

Thus for example, the degree of speed acquired by the moveable in the descent of four braccia would be double that which it had after falling through the space of two, and this would be the double of that resulting in the space of the first braccio. For there seems to me to be no doubt that the heavy body coming from a height of six braccia has, and strikes with, double the impetus that it would have from falling three braccia, and triple that which it would have from two, and six times that had in the space of one.

SALV. It is very comforting to have had such a companion in error, and I can tell you that your reasoning has in it so much of the plausible and probable, that our Author himself did not deny to me, when I proposed it to him, that he had labored for some time under the same fallacy. But what made me marvel then was to see revealed, in a few simple words, to be not only false but impossible, two propositions which are so plausible that I have propounded them to many people, and have not found one who did not freely concede them to me.

SIMP. Truly, I should be one of those who concede them. That the falling heavy body vires acquirat eundo [acquires force in going], the speed increasing in the ratio of the space, while the momentum of the same percussent is double when it comes from double height, appear to me as propositions to be granted without repugnance or controversy.

SALV. And yet they are as false and impossible as [it is] that motion should be made instantaneously, and here is a very clear proof of it. When speeds have the same ratio as the spaces passed or to be passed, those spaces come to be passed in equal times; if therefore the speeds with which the falling body passed the space of four braccia were the doubles of the speeds with which it passed the first two braccia, as one space is double the other space, then the times of those passages are equal; but for the same moveable to pass the four braccia and the two in the same time cannot take place except in instantaneous motion. But we see that the falling heavy body makes its motion in time, and passes the two braccia in less [time] than the four; therefore it is false that its speed increases as the space.

The other proposition is shown to be false with the same clarity. For that which strikes being the same body, the difference and momenta of the impacts must be determined only by the difference of the speeds; if therefore the percussent coming from a double height delivers a blow of double momentum, it must strike with double speed; but double speed passes the double space in the same time, and we see the time of descent to be longer from the greater height.»[16]

Salviati does not tell how Galileo has got the relation $v = at$ (we write it in present terms for short), but, by way of compensation, tries to explain to Simplicio in the subsequent steps for what reason the relation $v = as$, he held in the past, is wrong.

It's a pity that this easy-going explanation has stirred up generations of historians of science intent on pointing out how it is wrong as well. On this subject, during the last 120 years, an endless bibliography has piled up. We do not try to list it, we limit ourselves to tell that starting from the work of Raffaello Caverni,[17] of E. Wohlwill,[18] E. Mach,[19] P. Duhem,[20] A. Koyré,[21] over the more recent W. Wisan, Stillman Drake, one arrives at present-day.

The argument has been tackled from different points of view, i.e. both mere calculation and methodology or historical-critical survey.

For instance, B. Cohen[22] has shown that Galileo's explanation is based on a wrong application of the Mertonian rule (which enables to report a uniformly accelerated motion to a rectilinear uniform motion, as we have already recalled) to each instant Δt of the motion.

Before that, Mach, by making use of infinitesimal calculus, easily demonstrated the inconsistency of the relation $v = as$. In fact, obviously, the controversial relation must be written as $\frac{ds}{dt} = as$ which, once integrated, gives $s = Ke^{at}$, where K is an arbitrary integration constant and a the constant acceleration. At the initial

[16]E. N. VIII, pp. 202–204; Drake, pp. 158–161.

[17]Raffaello Caverni: **Storia del Metodo Sperimentale in Italia** 1891–1900—Tomo IV, p. 295 and ff.

[18]E. Wohlwill: **Die Entdeckung der Parabelform der Wurflinie**—Abhandlungen zur Geschichte der Mathematik, 577–624, 1899 (See an English translation in Science in Context 13, 375–410, 2000).

[19]E. Mach: **The Science of Mechanics: A Critical and Historical account of its Development** La Salle; III; The open court publishing, 1960 (translation of the 9. German edition).

[20]P. Duhem: **Études sur Léonard de Vinci**, troisième série, 1913—pp. 562–583.

[21]A. Koyré: **La loi de la chute des corps: Descartes et Galilée**—Paris: Herman, 1939 (pp. 76–150).

[22]B. Cohen: **Galileo's Rejection of the Possibility of Velocity Changing Uniformly with Respect to Distance**—Isis, 47, 231–235, 1956.

time $t = 0$, it will be $s = 0$ as well, so $K = 0$ and the relation inconsistent, i.e. the motion cannot even start.

Cohen reminds that Mach couldn't do otherwise since, at that time, he didn't knew the studies on the medieval dynamics (and thus the existence of the Mertonian rule). Actually, in Cohen's opinion, the explanation given by Galileo for his wrong choice had a historical and contextual origin.

As a confirmation of this, before Cohen, Koyré involved in the historical survey Descartes as well, "guilty" of the same error made by Galileo, even if for different reasons.

Finally, Stillman Drake,[23] by discussing the interpretations given by the scholars who used translations of the **Discourses** (in English, French and German), draws the attention (in particular) to the most widespread English translation.[24] In Drake's opinion the origin of wrong conclusions is in the mistakes of these translations. Obviously, in turn, Drake proposes his own conclusions which will be later discussed and, in part, disputed by M.A. Finocchiaro.[25]

What one can say in conclusion is that, in spite of a painstaking study, mainly by Drake, of the Galilean manuscripts not published in the national edition (Favaro), a truly verifiable evidence of when and where Galileo shifted from the hypothesis $v = as$ to the ultimate $v = at$ has not been found.

In 1975, Drake (by interpreting an unpublished manuscript of Galileo—folio 107^v in volume 72 of Galilean manuscripts of National Library of Florence) has conjectured that Galileo has performed an experiment in which he could measure times smaller than a second, making use of his musical skills. This experiment, never explicitly mentioned by Galileo, would have originated the discovery of the law of fall.[26] Drake's hypothesis has been discussed and criticized by R.H. Naylor, who jointly with Th. B. Settle tried to repeat that experiment.[27]

The only thing we know for sure is that the famous letter to Paolo Sarpi[28] of 1604 proves that at that time Galileo knew the exact law of fall ($s = \frac{1}{2}at^2$), maybe derived from experiments, but was still convinced of the wrong relation $v = as$.

After having exchanged a few words with Sagredo, at last Salviati concludes:

«….. the Author requires and takes as true one single assumption; that is:

---

[23]Stillman Drake: **Galileo studies**—The University of Michigan Press, 1970. pp. 229–239, **Uniform Acceleration, Space and Time**—British Journal for History of Science, 5, 28–43, 1970.

[24]Galileo Galilei: **Dialogues concerning Two New Sciences**—Trans. by Henry Crew and Alfonso de Salvio (Original edition 1914)—Dover, 1951.

[25]Maurice A. Finocchiaro: **Vires Acquirit Eundo—The passage where Galileo renounces space-acceleration and causal investigation**—Physis XIV, 125–146, 1972.

[26]S. Drake: **Galileo's work on free fall in 1604** Physis, 16, 309–322, 1974; **The Role of Music in Galileo's Experiments**—Scientific American, 232, 98–104, 1975.

[27]R. H. Naylor: **The Role of Experiment in Galileo's Early Work on the Law of Fall**—Annals of Science, 37, 363–378, 1980.

[28]E. N. X, pp. 114–116. We quote in Appendix two excerpts (from the question of Sarpi and the answer of Galileo)—translation by Renn, Damerow and Rieger, op. cit. pp. 84–85.

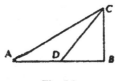

Fig. 5.2

[POSTULATE]

I assume that the degrees of speed acquired by the same moveable over different inclinations of planes are equal whenever the heights of those planes are equal.»[29]

Then, he continues by saying what the Author means by the height of an inclined plane:

«He calls the "height" of an inclined plane that vertical from the upper end of the plane which falls on the horizontal line extended through the lower end of the said inclined plane. For an understanding of this, take line *AB* parallel to the horizon (Fig. 5.2), upon which are the two inclined planes *CA* and *CD*; the vertical *CB*, falling to the horizontal *BA*, is called by the Author the height [or altitude, or elevation] of planes *CA* and *CD*.

Here he assumes that the degrees of speed of the same moveable, descending along the inclined planes *CA* and *CD* to points *A* and *D*, are equal, because their height is the same *CB*; and the like is also to be understood of the degree of speed that the same body falling from the point *C* would have at *B*.

SAGR. This assumption truly seems to me to be so probable as to be granted without argument, supposing always that all accidental and external impediments are removed, and that the planes are quite solid and smooth, and that the moveable is of perfectly round shape, so that both plane and moveable alike have no roughness.

With all obstacles and impediments removed, my good sense [*il lume naturale*] tells me without difficulty that a heavy and perfectly round ball, descending along the lines *CA*, *CD*, and *CB*, would arrive at the terminal points *A*, *D*, and *B* with equal impetus.

SALV. You reason from good probability. But apart from mere plausibility, I wish to increase the probability so much by an experiment that it will fall little short of equality with necessary demonstration.

---

[29]E. N. VIII, p. 205. Drake p. 162. If we read it in modern terms, we can say that it is a statement of the conservation of mechanical energy: the potential energy of the moveable at the peak of the inclined plane equals its kinetic energy when it reaches the ground, hence the ground speed, for any moveable, depends only on the height from which it started. Galileo, even if always within this limited ambit, already used this principle in the **Dialogue** (E. N. VII, p. 47). Notwithstanding one can find similar enunciations in Leonard and Cardan as well, and, implicitly, in the **De Ponderibus** of Jordanus de Nemore, if we read things in the actual perspective, the postulate (as Salviati calls it) provoked perplexities in several people at a point to oblige Galileo to justify it by an addition he dictated to Viviani. One must make clear that, this passage was added in the second Italian edition (Bologna 1655–56). We shall devote a part of the next section to an account of the matter.

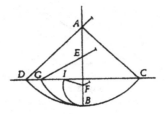

Fig. 5.3

Imagine this page to be a vertical wall, and that from a nail driven into it, a lead ball of one or two ounces hangs vertically, suspended by a fine thread two or three braccia in length, *AB* (Fig. 5.3). Draw on the wall a horizontal line *DC*, cutting at right angles the vertical *AB*, which hangs a couple of inches out from the wall; then, moving the thread *AB* with its ball to *AC*, set the ball free. It will be seen first to descend, describing the arc *CB*, and then to pass the point *B*, running along the arc *BD* and rising almost up to the parallel marked *CD*, falling short of this by a very small interval and being prevented from arriving there exactly by the impediment of the air and the thread.

From this we can truthfully conclude that the impetus acquired by the ball at point *B* in descent through arc *CB* was sufficient to drive it back up again to the same height through a similar arc *BD*. Having made and repeated this experiment several times, let us fix in the wall along the vertical *AB*, as at *E* or *F*, a nail extending out several inches, so that the thread *AC*, moving as before to carry the ball *C* through the arc *CB*, is stopped when it comes to *B* by this nail, *E*, and is constrained to travel along the circumference *BG*, described about the center *E*.

We shall see from this that the same impetus can be made that, when reached at *B* before, drove this same moveable through the arc *BD* to the height of horizontal *CD*, but now, gentlemen, you will be pleased to see that the ball is conducted to the horizontal at point *G*. And the same thing happens if the nail is placed lower down, as at *F*, whence the ball will describe the arc *BI*, ending its rise always precisely at the same line, *CD*. If the interfering nail is so low that the thread advancing under it could not get up to the height CD, as would happen when the nail was closer to point *B* than to the intersection of *AB* with the horizontal *CD*, then the thread will ride on the nail and wind itself around it.

This experiment leaves no room for doubt as to the truth of our assumption, for the two arcs *CB* and *DB* being equal and similarly situated, the acquisition of momentum made by descent through the arc *CB* is the same as that made by descent through the arc *DB*; but the momentum acquired at *B* through arc *CB* is able to drive the same moveable back up through arc *BD*, whence also the momentum acquired in the descent *DB* is equal to that which drives the same moveable through the same arc from *B* to *D*. So that in general, every momentum acquired by descent through an arc equals one which can make the same moveable rise through that same arc; and all the momenta that make it rise through all the arcs *BD*, *BG* and *BI* are equal, because they are created by the same momentum acquired through the descent *CB*, as experiment shows. Hence all the momenta acquired through descents along arcs *DB*, *GB*, and *IB* are equal.

SAGR. The argument appears to me conclusive, and the experiment is so well adapted to verify the postulate that it may very well be worthy of being conceded as if it had been proved.

SALV. I do not want any of us to assume more than need be, Sagredo; especially because we are going to make use of this assumption chiefly in motions made along straight surfaces, and not curved ones, in which acceleration proceeds by degrees very different from those that we assume it to take when it proceeds in straight lines.

Fig. 5.4

The experiment adduced thus shows us that descent through arc *CB* confers such momentum on the moveable as to reconduct it to the same height along any of the arcs *BD*, *BG*, or *BI*. But we cannot show on this evidence that the same would happen when [even] a most perfect sphere is to descend along straight planes inclined according to the tilt of the chords of those arcs. Indeed, we may believe that since straight planes would form angles at point *B*, a ball that had descended along the incline through the chord *CB* would encounter obstruction from planes ascending according to chords *BD*, *BG*, or *BI* and in striking against those, it would lose some of its impetus, so that in rising it could not get back to the height of line *CD*.

But if the obstacle that prejudices this experiment were removed, it seems to me that the mind understands that the impetus, which in fact takes [its] strength from the amount of the drop, would be able to carry the moveable back up to the same height.

Hence let us take this for the present as a postulate, of which the absolute truth will be later established for us by our seeing that other conclusions, built on this hypothesis, do indeed correspond with and exactly conform to experience.

This postulate alone having been assumed by the Author, he passes on to the propositions, proving them demonstratively[30]; and the first is this:».[31]

PROPOSITION I. THEOREM I

«*The time in which a certain space is traversed by a moveable in uniformly accelerated movement from rest is equal to the time in which the same space would be traversed by the same moveable carried in uniform motion whose degree of speed is one-half the maximum and final degree of speed of the previous, uniformly accelerated, motion.*

Let line AB (Fig. 5.4) represent the time in which the space CD is traversed by a moveable in uniformly accelerated movement from rest at C. Let EB, drawn in any way upon AB, represent the maximum and final degree of speed increased in the instants of the time AB.

---

[30]As we have seen, in the **Dialogue**, in this regard Sagredo asks «Is there a mathematical proof of this statement?» and Salviati answers «Most purely mathematical, ...».

[31]E. N. VIII, pp. 205–208; Drake, pp. 162–165.

All the lines reaching AE from single points of the line AB and drawn parallel to BE will represent the increasing degrees of speed after the instant A. Next, I bisect BE at F, and I draw FG and AG parallel to BA and RP; the parallelogram AGFB will [thus] be constructed, equal to the triangle AEB, its side GF bisecting AE at I.

Now if the parallels in triangle AEB are extended as far as IG, we shall have the aggregate of all parallels contained in the quadrilateral equal to the aggregate of those included in triangle AEB, for those in triangle IEF are matched by those contained in triangle GIA, while those which are in the trapezium AIFB are common.

Since each instant and all instants of time AB correspond to each point and all points of line AB, from which points the parallels drawn and included within triangle AEB represent increasing degrees of the increased speed, while the parallels contained within the parallelogram represent in the same way just as many degrees of speed not increased but equable, it appears that there are just as many momento, of speed consumed in the accelerated motion according to the increasing parallels of triangle AEB, as in the equable motion according to the parallels of the parallelogram GB.

For the deficit of momenta in the first half of the accelerated motion (the momenta represented by the parallels in triangle AGI falling short) is made up by the momenta represented by the parallels of triangle IEF.

It is therefore evident that equal spaces will be run through in the same time by two moveables, of which one is moved with a motion uniformly accelerated from rest, and the other with equable motion having a momentum one-half the momentum of the maximum speed of the accelerated motion; which was [the proposition] intended.»[32]

Summarizing:

1. Segment AB, consisting of an infinite set of points, represents the interval on the time axis corresponding to space CD (covered by uniformly accelerated motion). Hence, by reasoning in modern terms, the space is the independent variable.
2. The segments perpendicular to AB represent the velocities at each instant, EB the velocity reached by the uniformly accelerated motion at the instant in which the moveable reaches the point D, $FB = \frac{1}{2}EB$ the constant velocity of a hypothetical moveable which goes from C to D with uniform motion.
3. The area of triangle AEB represents the whole of all the "degrees of velocity" owned by the moveable in the uniformly accelerated motion from C to D. The area of rectangle ABFG represents the whole of the "degrees of velocity", constantly equal to EB, of the hypothetical uniform rectilinear motion.[33]
4. The equality of the areas of triangles AGI and EFI demonstrates the assertion.

If we really want to quibble, we can object that Galileo starts by saying that he wants to demonstrate the equality of the times of the two motions and, on the

---

[32]E. N. VIII, p. 208; Drake, pp. 165–166.

[33]Implicitly, Galileo considers an area constituted by the union of an infinity of segments, as if there were an infinite number of pot-hooks drawn one close to another. The influence of the contemporary work of Cavalieri and the study of «indivisibles» is evident, as first Caverni pointed out. (op. cit. p. 309).

contrary, he demonstrates (see the final phrase) the equality of the spaces (covered in the same time).

The same happens also when one wants to proceed by making use of the modern methods of calculation (see later).

Let us pass to

PROPOSITION II. THEOREM II

«If a moveable descends from rest in uniformly accelerated motion, the spaces run through in any times whatever are to each other as the duplicate ratio of their times; that is, are as the squares of those times.

Let the flow of time from some first instant A be represented by the line AB (Fig. 5.5), in which let there be taken any two times, AD and AE. Let HI be the line in which the uniformly accelerated moveable descends from point H as the first beginning of motion; let space HL be run through in the first time AD, and HM be the space through which it descends in time AE.

I say that space MH is to space HL in the duplicate ratio of time EA to time AD. Or let us say that spaces MH and HL have the some ratio as do the squares of EA and AD.

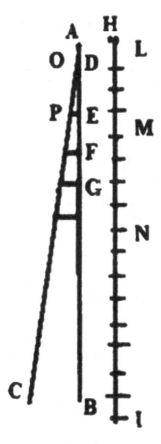

Fig. 5.5

*Draw line AC at any angle with AB. From points D and E draw the parallels DO and EP,
of which DO will represent the maximum degree of speed acquired at instant D of time
AD, and PE the maximum degree of speed acquired at instant E of time AE.*

*Since it was demonstrated above that as to spaces run through, those are equal to one
another of which one is traversed by a moveable in uniformly accelerated motion from
rest, and the other is traversed in the same time by a moveable carried in equable motion
whose speed is one-half the maximum acquired in the accelerated motion, it follows that
spaces MH and LH are the same that would be traversed in times EA and DA in equable
motions whose speeds are as the halves of PE and OD.*

*Therefore if it is shown that these spaces MH and LH are in the duplicate ratio of the
times EA and DA, what is intended will be proved.*

*Now in Proposition IV of Book I it was demonstrated that the spaces run through by move-
ables carried in equable motion have to one another the ratio compounded from the ratio of
speeds and from the ratio of times. Here, indeed, the ratio of speeds is the same as the ratio
of times, since the ratio of one-half PE to one-half OD, or of PE to OD, is that of AE to AD.*

*Hence the ratio of spaces run through is the duplicate ratio of the times; which was to be
demonstrated.*

*It also follows from this that this same ratio of spaces is the duplicate ratio of the maxi-
mum degrees of speed; that is, of lines PE and OD, since PE is to OD as EA is to DA.*

COROLLARY I

*From this it is manifest that if there are any number of equal times taken successively from
the first instant or beginning of motion, say AD, DE, EF, and FG, in which spaces HL,
LM, MN, and NI are traversed, then these spaces will be to one another as are the odd
numbers from unity, that is, as 1, 3, 5, 7; but this is the rule [ratio] or excesses of squares
of lines equally exceeding one another [and] whose [common] excess is equal to the least
of the same lines, or, let us say, of the squares successively from unity, Thus when the
degrees of speed are increased in equal times according to the simple series of natural
numbers, the spaces run through in the same times undergo increases according with the
series of odd numbers from unity.»*[34]

«SAGR. Please suspend the reading for a bit, while I develop a fancy that has come to
my mind about a certain conception. To explain this, and for my own as well as for your
clearer understanding. I'll draw a little diagram (Fig. 5.6). I imagine by this line *AI* the
progress of time after the first instant at *A*; and going from *A* at any angle you wish, I draw
the straight line *AF*. And joining points and F, I divide the time *AI* at the middle in C, and
I draw *CB* parallel to *IF*, taking *CB* to be the maximum degree of the speed which, com-
mencing from rest at *A*, grows according to the increase of the parallels to *BC* extended in
triangle *ABC*; which is the same as to increase [according] as the time increases.

I assume without argument, from the discussion up to this point, that the space passed
by the moveable falling with its speed increased in the said way is equal to the space that
would be passed by the same moveable if it were moved during the same time *AC* in uniform motion whose degree of speed was equal to *EC*, one-half of *BC*.

I now go on to imagine the moveable [to have] descended with accelerated motion and
to be found at instant *C* to have the degree of speed *BC*. It is manifest that if it continued
to be moved with the same degree of speed *BC*, without accelerating further, then in the

---

[34]E. N. VIII, pp. 209–210; Drake, pp. 166–167.

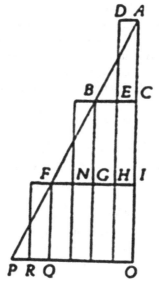

Fig. 5.6

ensuing time CI it, would pass a space double that which it passed in the equal time AC with degree of uniform speed EC, one-half the degree BC.

But since the moveable descends with speed always uniformly increased in all equal times, it will add to the degree CB, in the ensuing time CI, those same momenta of speed growing according to the parallels of triangle BFG, equal to triangle ABC; so that to the degree of speed G there being added one-half the degree FG, the maximum of those [speeds] acquired in the accelerated motion governed by the parallels of triangle BFG, we shall have the degree of speed IN, with which it would be moved with uniform motion during time CI.

That degree IN is triple the degree EC convinces [us] that the space passed in the second time CI must be triple that [which was] passed in the first time CA.

And if we assume added to AI a further equal part of time IO, and enlarge the triangle out to A PO, then it is manifest that if the motion continued through the whole time IO with the degree of speed IF acquired in the accelerated motion during time AI, this degree IF being quadruple EC, the space passed in time IO would be quadruple that passed in the first equal time AC. Continuing the growth of uniform acceleration in triangle FPQ, similar to that of triangle ABC which, reduced to equable motion, adds the degree equal to EC, and adding QR equal to EC, we shall have the entire equable speed exercised over time IO quintuple the equable [speed] of the first time AC; and hence the space passed [will be] quintuple that [which was] passed in the first time AC.

Thus you see also, in this simple calculation, that the spaces passed in equal times by a moveable which, parting from rest, acquires speed in agreement with the growth of time, are to one another as the odd numbers from unity, 1,3,5; and taking jointly the spaces passed, that which is passed in double the time is four times that passed in the half [i.e., in the given time], and that passed in triple the time is nine times [as great]. And in short, the spaces passed are in the duplicate ratio of the times; that is, are as the squares of those times.»[35]

---

[35]E. N. VIII, pp. 210–212; Drake, pp. 167–169.

Before continuing our talk, let us try to turn in our usual terms, that is those of the Newtonian mechanics, the two propositions enunciated, and demonstrated by Galileo. At present, the relations regarding the uniformly accelerated motion are obtained by successively integrating $\frac{d^2s}{dt^2} = a = const.$, where s is the space covered, t the time, and a the constant acceleration. With the initial conditions $s(t = 0) = 0$, $v(t = 0) = 0$, where v indicates the velocity, we shall obtain

$$v = at, \quad s = \frac{1}{2}at^2.$$

As regards proposition I, let us see how to verify it. Supposing it is true. If the space covered at time t is $s = \frac{1}{2}at^2$ with final velocity $v = at$, the space covered in the same time with uniform motion and velocity $\frac{1}{2}v$ will be as well $s = \frac{1}{2}v \cdot t = \frac{1}{2}at \cdot t = \frac{1}{2}at^2$.

We shall come back later on the formal coincidence of this proposition with the "Merton College rule". As regards proposition II and the series of the odd numbers starting from unity, let us start once again from $s = \frac{1}{2}at^2$. If we consider the spaces covered, starting from rest, at the subsequent time units, we shall have

$$s_1 = \frac{1}{2}a.1$$

$$s_2 = \frac{1}{2}a.4$$

$$s_3 = \frac{1}{2}a.9$$

$$s_4 = \frac{1}{2}a.16$$

$$s_5 = \frac{1}{2}a.25$$

$$\dots$$

Hence, the factor $\frac{1}{2}a$ apart, the subsequent intervals are

$$s_1 = 1 - 0 = 1$$
$$s_2 - s_1 = 4 - 1 = 3$$
$$s_3 - s_2 = 9 - 4 = 5$$
$$s_4 - s_3 = 16 - 9 = 7$$
$$s_5 - s_4 = 25 - 16 = 9$$
$$\dots$$

Which is exactly the series of odd numbers starting from unity.

5.3.1 Appendix—From the Correspondence of Paolo Sarpi and Galileo

Paolo Sarpi wrote (from Venice) to Galileo (in Padua) the 9th of October 1604:

«We have already concluded that a body cannot be thrown up to the same point [termine] if not by a force, and, accordingly, by a velocity. We have recapitulated - so Your Lordship lately argued and originally found out [inventòella] - that [the body] will return downwards through the same points through which it went up. There was, I do not remember precisely [non so che], an objection concerning the ball of the arquebus; in this case, the presence of the fire troubles the strength of the argument. Yet, we say: a strong arm which shoots an arrow with a Turkish bow completely pierces through a table; and when the arrow descends from that height to which the arm with the bow can take it, it will pierce [the table] only slightly. I think that the argument is maybe slight, but I do not know what to say about it.»[36]

The 16th of October Galileo answered:

«Thinking again about the matters of motion, in which, to demonstrate the phenomena [accidenti] observed by me, I lacked a completely indubitable principle which I could pose as an axiom, I am reduced to a proposition which has much of the natural and the evident: and with this assumed, I then demonstrate the rest; i.e., that the spaces passed by natural motion are in double proportion to the times, and consequently the spaces passed in equal times are as the odd numbers from one, and the other things. And the principle is this: that the natural moveable goes increasing in velocity with that proportion with which it departs from the beginning of its motion; (…)».[37]

5.4 The Historical Experiment and the Postulate

From the excerpts quoted above we have the last version, in order of time (1638), of the Galilean mechanics of the uniformly accelerated motion and the law of fall. We have omitted the excerpts regarding the rectilinear uniform motion since Galileo arranges in rigorous mathematical form (even if not corresponding to the present form) results which were already well-known. As Enrico Giusti says

«… it's a matter of a non-controversial theory: unlike what will happen for the uniformly accelerated motion, in this case the principles are accepted by everyone, the results are common knowledge, the demonstrations impeccable.»[38]

In addition to the treatment of the uniformly accelerated motion, Galileo (in the part of the dialogue between Simplicio and Salviati we report below) maintains that the fall of bodies takes place with uniformly accelerated motion.

---

[36]E. N. X, p. 114; (J. Renn et al. p. 84).

[37]E. N. X, p. 115; (J. Renn et al. pp. 84–85).

[38]See in this regard: E. Giusti- **Ricerche galileiane: il trattato De motu equabili come modello della teoria delle proporzioni** (Bollettino di Storia delle Scienze Matematiche, VI-2, 89, 1986).

«SIMP. Really I have taken more pleasure from this simple and clear reasoning of Sagredo's than from the (for me) more obscure demonstration of the Author, so that I am better able to see why the matter must proceed in this way, once the definition of uniformly accelerated motion has been postulated and accepted. But I am still doubtful whether this is the acceleration employed by nature in the motion of her falling heavy bodies. Hence, for my understanding and for that of other people like me, I think that it would be suitable at this place [for you] to adduce some experiment from those (of which you have said that there are many) that agree in various cases with the demonstrated conclusions.

SALV. Like a true scientist, you make a very reasonable demand, for this is usual and necessary in those sciences which apply mathematical demonstrations to physical conclusions, as may be seen among writers on optics, astronomers, mechanics, musicians, and others who confirm their principles with sensory experiences, those being foundations of all the resulting structure. I do not want to have it appear a waste of time *[superfluo]* on our part, [as] if we had reasoned at excessive length about this first and chief foundation upon which rests an immense framework of infinitely many conclusions of which we have only a tiny part put down in this book by the Author, who will have gone far to open the entrance and portal that has until now been closed to speculative minds. Therefore as to the experiments: the Author has not failed to make them, and in order to be assured that the acceleration of heavy bodies falling naturally does follow the ratio expounded above, I have often made the test *[prova]* in the following manner, and in his company.

In a wooden beam or rafter about twelve braccia long, half a braccio wide, and three inches thick, a channel was rabbeted in along the narrowest dimension, a little over an inch wide and made very straight; so that this would be clean and smooth, there was glued within it a piece of vellum, as much smoothed and cleaned as possible.

In this there was made to descend a very hard bronze ball, well rounded and polished, the beam having been tilted by elevating one end of it above the horizontal plane from one to two braccia, at will. As I said, the ball was allowed to descend along *[per]* the said groove, and we noted (in the manner I shall presently tell you) the time that it consumed in running all the way, repeating the same process many times, in order to be quite sure as to the amount of time, in which we never found a difference of even the tenth part of a pulse-beat.

This operation being precisely established, we made the same ball descend only one-quarter the length of this channel, and the time of its descent being measured, this was found always to be precisely one-half the other. Next making the experiment for other lengths, examining now the time for the whole length [in comparison] with the time of one-half, or with that of two-thirds, or of three-quarters, and finally with any other division, by experiments repeated a full hundred times, the spaces were always found to be to one another as the squares of the times.

And this [held] for all inclinations of the plane; that is, of the channel in which the ball was made to descend, where we observed also that the times of descent for diverse inclinations maintained among themselves accurately that ratio that we shall find later assigned and demonstrated by our Author.

As to the measure of time, we had a large pail filled with water and fastened from above, which had a slender tube affixed to its bottom, through which a narrow thread of water ran; this was received in a little beaker during the entire time that the ball descended along the channel or parts of it.

The little amounts of water collected in this way were weighed from time to time on a delicate balance, the differences and ratios of the weights giving us the differences and ratios of the times, and with such precision that, as I have said, these operations repeated time and again never differed by any notable amount.

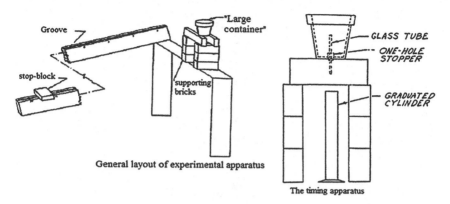

General layout of experimental apparatus

The timing apparatus

Fig. 5.7

SIMP. It would have given me great satisfaction to have been present at these experiments. But being certain of your diligence in making them and your fidelity in relating them, I am content to assume them as most certain and true.»[39]

But, in the course of time, not everybody agreed with Simplicio. Putting aside the criticisms of the contemporaneous Descartes, let's move to XX century and to the man who more has questioned the "experimental" results of Galileo, that is, Alexandre Koiré. In fact, on the subject of the experiment described above, he comments:

«The experiment which Galileo performs is wonderfully devised; the idea of replacing the free fall with the fall on an inclined plane is, truly, ingenious. But it is true as well that the execution is not up to the work of the idea … Luckily, since otherwise nobody would have imagined a so rigorous correspondence of experiment with the forecasts, on the contrary, despite Galileo's claim one is tempted to doubt it, for the simple reason that such a rigorous correspondence is *strictly impossible*.»[40]

And this is not the only occasion in which he accuses Galileo of talking of experiments really performed whereas on the contrary they would have been "ideal experiments".

Unlike Koiré, there has been a person who has believed in Galileo's skill as experimenter.

This was Thomas B. Settle,[41] who made the same experiment again, with the same means, and thus confirmed, not only the feasibility and the suitability of the equipment, but also the accuracy reached by Galileo.

We reproduce here the figures (see Fig. 5.7) by which Settle represents his experimental apparatus. Notwithstanding the evidence represented by Settle's

---

[39]E. N. VIII, pp. 212–213; Drake pp. 169–170.

[40]A. Koyré: **La loi de la chute des corps**, op. cit. pp. 71–72 (our translation).

[41]See: Thomas B. Settle—**An Experiment in the History of Science**—Science, 133, 19–23, 1961.

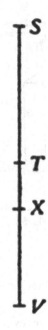

Fig. 5.8

experiment, Koyré's opinion continued to be maintained in literature for years, as documented by Stillman Drake.[42]

Let us leave now the historic experiment and come back to the theory for dealing with the "postulate". But, following Salviati, it is necessary to place before the corollary and the subsequent scholium:

COROLLARY II

It is deduced, second, that if at the beginning of motion there are taken any two spaces whatever, run through in any [two] times, the times will be to each other as either of these two spaces is to the mean proportional space between the two given spaces.

From the beginning of motion, S, (Fig. 5.8) take two spaces, ST and SV, of which the mean proportional shall be SX; the time of fall through ST will be to the time of fall through SV as ST is to SX; or let us say that the time through SV is to the time through ST as VS is to SX. Since it has-been demonstrated that the spaces run through are in the duplicate ratio of the times (or what is the same thing, are as the squares of the times), the ratio of space VS to space ST is the doubled ratio of VS to SX, or is the same as that of the squares of VS and SX. It follows that the ratio of times of motion through SV and ST are as the spaces, or the lines, VS and SX.

SCHOLIUM

What we have demonstrated for movements run through along verticals is to be understood also to apply to planes, however inclined; for these, it is indeed assumed that the

---

[42]S. Drake: **Galileo's experimental configurations of horizontal inertia—Unpublished manuscripts**—Isis, 64, 290–305, 1973.

*degree of increased speed [accelerationis] grows in the same ratio; that is, according to
the increase of time, or let us say according to the series of natural numbers from unity.»*[43]

At this point there is still a question to be remarked regarding the constancy of
the gravity acceleration.

It is obvious that, for Galileo, all refers to bodies which fall from a certain
height, exceedingly smaller than the Earth's radius. Therefore, as we do today as
well (neglecting the exact Newtonian expression), it is more than legitimate to
assume that (for an inertial observer) all bodies, independently of their mass, fall
with the same acceleration g (measured at ground level).

At last, let us speak about the "postulate".

We have seen that Salviati, after having substantiated the equality of velocities
of bodies, which arrive at ground falling from inclined planes of different inclination but of the same height, suggests to accept this equality as a postulate. In
fact the justification supplied resorts to an example when the trajectories are not
rectilinear and therefore he remains on the safe side .... I do not want any of us to
assume more than need be, Sagredo; ....

In effect it was a matter of the greatest importance.

As we have already remarked, that assertion—if read with today's judgment—
means the conservation of mechanical energy (even if enunciated only in relation
to the motion on inclined planes).

Obviously, Galileo couldn't attach it the role of a fundamental conservation
law that it plays today, but the fact that this law is often recalled in his work, also
before the **Discourses** (for instance, repeatedly in the first Day of the **Dialogue**),
as if he would claim the ownership of it in any case, cannot pass unnoticed.

The acceptance of the "postulate" by the readers (of the 1638 Elzevirian edition) was not immediate and unquestioned. A part of the scientific community of
that time (in Italy mostly made-up of clergymen of various orders) was in touch
(often by letter) with Galileo. One of the few not ecclesiastic, the Genoese patri-
cian Giovan Battista Baliani who in the past had already corresponded with
Galileo (Salviati was the person who put them in contact) wrote to Galileo the first
of July 1639, answering a letter of him (of the twentieth of June) which accompanied the forwarding of a copy of the **Discourses**. After the usual courteousness in
the emphatic forms customary at that time, Baliani just poses a question regarding
the "postulate".[44] Before reporting the relevant passages of Galileo's reply, we
have to add some words on the author of the letter.

Baliani, just the preceding year, had published a little work,[45] more or less on
the same matter dealt with in the **Discourses** (which in Italy were taking a long

---

[43]E. N. VIII, p. 214; Drake pp. 170–171.

[44]E. N. XVIII, pp. 68–71.

[45]**De Motu naturali gravium solidorum et liquidorum**—Genuae, 1638. In 1646, with the same
publisher, Baliani published a second edition of his work, now in six books and where the first
book was a remaking of the preceding booklet. On all this, see Giovan Battista Baliani—**De
Motu naturali gravium solidorum et liquidorum**—edited by Giovanna Baroncelli, Giunti,
1998 (in Italian, since an English translation does not exist).

time to be spread). What's more, the contents of the little work reported conclusions which, at least mostly, were in accordance with the results of Galileo.

Here are the passages of Galileo's reply we are interested in:

«I shall defer it to another time to satisfy that part [of your letter], full of generosity, and shall say now only, and briefly, something about the scientific particulars you touch on.

Then that the principle I assume on p. 166 [the postulate, mentioned earlier] does not, as you note, appear with that [self-]evidence that is required of principles to be postulated as known, I concede this to you now-though you make the same assumption yourself, that is, that the degrees of speed acquired by moveables descending by different planes through the same height are equal.

Know, then, that after my having lost my sight, and consequently my faculty of going more deeply into propositions and demonstrate more profound than those last discovered and written by me, I [instead] spent the nocturnal hours ruminating on the first and simplest propositions, reordering these and arranging them in better form and evidence. Among these it occurred to me to demonstrate the said postulate in the manner you will in time see, if I shall have sufficient strength to improve and amplify what was written and published by me up to now about motion by adding some little speculations, and in particular those relating to the force of percussion, in the investigation of which I have consumed hundreds and thousands of hours, and have finally reduced this to very easy explanation, so that people can understand it in less than half an hour of time.»[46]

The discussion between the two continued a little more, also on other arguments, but it would be beyond our aim to continue following it.

At last, the third of December of the same year, Galileo wrote to Benedetto Castelli a letter in which he manifests his dissatisfaction with the explanation of the postulate given in the **Discourses** and informs him that he is working out a new demonstration of it in collaboration with a young man, at that time his guest and disciple.[47]

The letter, in friendly terms, is signed Galileo Galilei Linceo cieco (blind).[48]

The young man to whom Galileo alluded was Vincenzio Viviani, who came in October of the same year 1639 to live in Arcetri together with him, to play the role of an assistant and scribe. He will remain in Arcetri till the death of the master in 1642.

The young Viviani was enough expert to face discussions with Galileo on the arguments dealt in the **Discourses** and therefore (as after all the master tells) he has certainly contributed to the solution of the problem besides the writing of the pages we shall now see.

«SALV. Here, Sagredo, I want permission to defer the present reading for a time, though perhaps I shall bore Simplicio, in order that I may explain further what has been said and proved up to this point. At the same time it occurs to me that, by telling you of some

---

[46]E. N. XVIII, pp. 75–79 (Translation by Stillman Drake—**Galileo at work** pp. 399–400).

[47]As we already said in Footnote 24, the new "demonstration" was inserted in the second edition of the **Discourses** in the **Opere di Galileo Linceo** …. In Bologna by the H.H. of Dozza, 1656. It corresponds to pp. 214–219 of the national edition (Favaro).

[48]E. N. XVIII, pp. 125–126.

mechanical conclusions reached long ago by our Academician, I can add new confirmation of the truth of that principle which has already been examined by us with probable reasonings and by experiments. More important, this will be geometrically proved after the prior demonstration of a single lemma that is elementary in the study of impetuses.

SAGR When you promise such gains, there is no amount of time I should not willingly spend in trying to confirm and completely establish these sciences of motion. For my part, I not only grant permission to you to satisfy us on this matter, but I even beg you to allay as swiftly as possible the curiosity you have aroused in me. I think Simplicio feels the same way about this.

SIMP. How can I say otherwise?

SALV. Then, since you give me leave, consider it in the first place as a well-known effect that the momenta or speeds of the same moveable are different on diverse inclined planes, and that the greatest [speed] is along the vertical. The speed diminishes along other inclines according as they depart more from the vertical and are more obliquely tilted. Whence the impetus, power *[talento]*, energy, or let us say momentum of descent, comes to be reduced in the underlying plane on which the moveable is supported and descends.

The better to explain this, let the line *AB* (Fig. 5.9) be assumed to be erected vertically on the horizontal *AC*, and then let it be tilted at different inclinations with respect to the horizontal, as at *AD*, *AE*, *AF*, etc. I say that the impetus of the heavy body for descending is maximal and total along the vertical *BA*, is less than that along *DA*, still less along *EA*, successively diminishes along the more inclined *FA*, and is finally completely extinguished on the horizontal *C A*, where the moveable is found to be indifferent to motion and to rest, and has in itself no inclination to move in any direction, nor yet any resistance to being moved.

Thus it is impossible that a heavy body (or combination thereof) should naturally move upward, departing from the common center toward which all heavy bodies mutually converge *[conspirano]*; and hence it is impossible that these be moved spontaneously except with that motion by which their own center of gravity approaches the said common center. Whence, on the horizontal, which here means a surface [everywhere] equidistant from the said [common] center, and therefore quite devoid of tilt, the impetus or momentum of the moveable will be null.

This change of impetus assumed, I must next explain something that our Academician, in an old treatise on mechanics written at Padua for the use of his pupils, demonstrated at length and conclusively in connection with his treatment of the origin and character of that marvelous instrument, the screw; namely, the ratio in which this change of impetus along planes of different inclinations takes place. Given the inclined plane *AF*, for

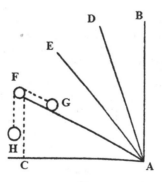

Fig. 5.9

example, and taking as its elevation above the horizontal the line FC, along which the impetus of a heavy body and its momentum in descent is maximum, we seek the ratio that this momentum has to the momentum of the same moveable along the incline *FA*, which ratio, I say, is inverse to that of the said lengths.

This is the lemma to be put before the theorem that I hope then to be able to demonstrate.

It is manifest that the impetus of descent of a heavy body is as great as the minimum resistance or force that suffices to fix it and hold it [at rest]. I shall use the heaviness of another moveable for that force and resistance, and [as] a measure thereof. Let the moveable G, then, be placed on plane *FA*, tied with a thread which rides over *F* and is attached to the weight *H*; and let us consider that the space of the vertical descent or rise of this *[H]* is always equal to the whole rise or descent of the other moveable, G, along the incline *AF* not just to the vertical rise or fall, through which the moveable G (like any other moveable) exclusively exercises its resistance. That much is evident. For consider the motion of the moveable G in the triangle *AFC* (for example, upward from *A* to *F*) as composed of the horizontal transversal *AC* and the vertical *CF*. As before, there is no resistance to its being moved along the horizontal, since by means of such a motion no loss or gain whatever is made with regard to its distance from the common center of heavy things, that being conserved always the same on the horizontal [as defined above]. It follows that the resistance is only with respect to compulsion to go up the vertical *CF*.

Hence the heavy body G, moving from *A* to F, resists in rising only the vertical space *CF*; but that other heavy body *H* must descend vertically as much as the whole space *FA*. And this ratio of ascent and descent remains always the same, being as little or as great as the motion of the said moveables by reason of their connection together.

Thus we may assert and affirm that when equilibrium (that is, rest) is to prevail between two moveables, their [overall] speeds or their propensions to motion that is, the spaces they would pass in the same time must be inverse to their weights *[gravità]*, exactly as is demonstrated in all cases of mechanical movements.

Thus, in order to hinder the descent of G, it will suffice that *H* be as much lighter than *G* as the space *CF* is proportionately less than the space *FA*.

Hence if the heavy body *G* is, to the heavy body H, as *FA* is to FC, equilibrium will follow; that is, the heavy bodies *H* and *G* will be of equal moments, and the motion of these moveables will cease.

Now, we have agreed that the impetus, energy, momentum, or propensity to motion of a moveable is as much as the minimum force or resistance that suffices to stop it; and it has been concluded that the heavy body *H* suffices to prohibit motion to the heavy body G; hence the lesser weight, which exercises its total [static] moment in the vertical FC, will be the precise measure of the partial moment that the greater weight G exercises along the inclined plane *FA*. But the measure of the total moment of heavy body G is G itself, since to hinder the vertical descent of a heavy body, there is required the opposition of one equally heavy when both are free to move vertically.

Therefore the partial impetus or momentum of G along the incline *FA* will be, to the maximum and total impetus of *G* along the vertical *FC*, as the weight *H* is to the weight G, which is (by construction) as the vertical *FC* (the height of the incline) is to the incline *FA* itself. This is what was proposed to be demonstrated as the lemma; and as we shall see, it is assumed by our Author as known in the second part of Proposition VI of the present treatise.

SAGR. It seems to me that from what you have concluded thus far, it can be easily deduced, arguing by perturbed equidistance of ratios, that the momenta of the same moveable along differently inclined planes having the same height, such as *FA* and *FI*, are in the inverse ratio of those same planes.

SALV. A true conclusion. This established, I go on next to demonstrate the theorem itself; that is :

[ADDED THEOREM]

The degrees of speed acquired by a moveable in descent with natural motion from the same height, along planes inclined in any way whatever, are equal upon their arrival at the horizontal, all impediments being removed.

Here you must first note that it has already been established that along any inclinations, the moveable upon its departure from rest increases its speed, or amount of impetus, in proportion to the time, in accordance with the definition given by the Author for naturally accelerated motion. Whence, as he has demonstrated in the last preceding proposition, the spaces passed are in the squared ratio of the times, and consequently of the degrees of speed.

Whatever the [ratio of) impetuses at the beginning *[nella prima mossa]*, that proportionality will hold for the degrees of the speeds gained during the same time, since both [impetuses and speeds] increase in the same ratio during the same time.

Now let the height of the inclined plane *AB* (Fig. 5.10) above the horizontal be the vertical *A C*, the horizontal being *CB*. Since, as we concluded earlier, the impetus of a moveable along the vertical *A C* is, to its impetus along the incline *AB*, as *AB* is to *AC*, [then] in the incline *AB* take *AD* as the third proportional of *AB* and *AC*, the impetus [to move] along *AC* is, to the impetus [to move] along *AB* (that is, [to move] along *AD*), as [*AB* is to *AC* or as] *AC* is to *AD*. Hence the moveable, in the same time that it passes the vertical space *AC*, would also pass the space *AD* along the incline *AB* (the momenta being as the spaces); and the degree of speed at *C* will have to' the degree of speed at *D* the same ratio that *AC* has to *AD*. But the speed at *B* is to the speed at *D* as the time through *AB* is to the time through *AD*, by our definition of accelerated motion; and the time through *AB* is to the time through *AD* as *AC* (the mean proportional between *BA* and *AD*) is to AD, by the last corollary to Proposition II. Therefore the speeds at *B* and C [both] have to the speed at *D* the same ratio that *AC* has to *AD*, and hence [the speeds at *B* and *C*] are equal; which is the theorem intended to be demonstrated.

From this we may more conclusively prove the Author's ensuing Proposition III, in which he makes use of the [earlier] postulate; this [theorem] states that the time along the incline has to the time along the vertical the same ratio that the incline has to the vertical. So let us say: If *EA* is the time along *AB*, the time along *AD* will be the mean proportional between these [*AB* and *AD*], that is, *AC*, by the second corollary to Proposition II. But if *AC* is the time along *AD*, it will also be the time along *AC*, since *AD* and *AC* are run through in equal times. And since if *EA* is the time along *AB*, *A C* will be the time along *AC*, then it follows that as *AB* is to *AC*, so is the time along *AB* to the time along *AC*.

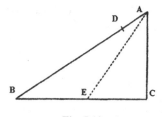

Fig. 5.10

By the same reasoning it will be proved that the time along AC is to the time along some other incline, AE, as AC is to AE, by equidistance of ratios, the time along incline AE is to the time along incline AE homologously as AB is to AE, etc.

As Sagredo will readily see [later], the Author's Proposition VI could be immediately proved from the same application of this theorem. But enough for now of this digression, which has perhaps turned out to be too tedious, though it is certainly profitable in these matters of motion.

SAGR. And not only greatly to my taste, but most essential to a complete understanding of that principle.

SALV. Then I shall resume the reading of the text.»[49]

The opening of Salviati rests on a circumstance accepted by the common sense, that the descent of a moveable (for instance a spherule) on inclined planes of different inclinations is faster when the inclination is greater up to the maximum of the rapidity when the inclination is of 90° ("a well-known effect").

In Fig. 5.9, "lines" AB, AD, AE, AF, are of equal length and different inclination above the horizon, so extremes B, D, E, F, are at different heights. We would say that the "well-known effect" can be put in quantitative terms by saying that (equal) distances are covered in different times according to the inclinations and thus we have the smallest time for the vertical motion.

Galileo says that the impetus of the heavy body in descending is maximum and total along perpendicular BA.

Further on the impetus is defined as "the impetus of descent of a heavy body is as great as the minimum resistance or force that suffices to fix it and hold it".

Hence, speaking in modern terms, in this case the impetus of a heavy body on an inclined plane is made like the weight multiplied by the sine of the inclination angle (in fact Galileo too, further on, says "I shall use the heaviness of another moveable for that force and resistance, and a measure thereof").

Then, he refers to what he has already exposed in **Le Mecaniche** to finally demonstrate the lemma we can enunciate as $\frac{I(FA)}{I(FC)} = \frac{FC}{FA}$, where I(FA) means the impetus for inclined line FA and I(FC) the impetus for vertical line FC. Obviously $\frac{FC}{FA}$ = sine of angle FAC.

At this point Galileo applies subsequently the two propositions already demonstrated, together with corollary II and the relevant scholium (which result to be fundamental for the demonstration). He passes from inclined planes of the same length and different inclinations (therefore of different heights) to planes of different inclinations but of the same height.

As a whole, the demonstration of the postulate is rather laborious so to have been defined "obscure" by Dijksterhuis[50] who noticed that the term impetus is used in two completely different meanings.[51]

---

[49]E. N. VIII, 214–219; Drake pp. 171–175.

[50]E. J. Dijksterhuis: **The Mechanization of the World Picture**, op. cit. pp. 347–348.

[51]The term "impetus", for Galileo, at first indicates the dynamic impulse supplied to a body by its gravity, but also, for the moveable at the end of its descent, the "impetus" takes the meaning of total energy. Again, for the moveable at rest, the "impetus" is the virtual velocity.

This fact happens also elsewhere and is obviously due to the non-existence of a true definition of velocity. It must be pointed out the fact that the equality of the velocities (at the arrival on the horizontal plane) of moveables falling along planes of different inclinations but of the same height, which we also read as conservation of the mechanical energy, is apparently demonstrated as a purely kinematic property.

Actually, a fundamental statement is that the heavy bodies fall with constant acceleration (that is the fall of heavy bodies is a uniformly accelerated motion).

Let us revise the argument in modern terms, by making use of the two relations relevant to the uniformly accelerated motions. By making reference to the figure, let us call $AC = l_1, AE = l_2, AB = h, A\hat{C}B = \vartheta_1, A\hat{E}B = \vartheta_2$.

Therefore, it will be $\sin\vartheta_1 = \frac{AB}{AC} = \frac{h}{l_1}, \sin\vartheta_2 = \frac{AB}{AE} = \frac{h}{l_2}$ and, by applying the two relations of the uniformly accelerated motion.

$$l_1 = \frac{1}{2}g\sin\vartheta_1 t_1^2 = \frac{1}{2}g\frac{h}{l_1}t_1^2,$$

$$l_2 = \frac{1}{2}g\sin\vartheta_2 t_2^2 = \frac{1}{2}g\frac{h}{l_2}t_2^2,$$

$$v_C = g\sin\vartheta_1 t_1 = g\frac{h}{l_1}t_1,$$

$$v_E = g\sin\vartheta_2 t_2 = g\frac{h}{l_2}t_2.$$

From which, by eliminating t_1 and t_2 (the times of arrival in C and in E) among the four relations, one obtain $\frac{v_C}{v_E} = 1 \Rightarrow v_C = v_E$.

The demonstration of the "postulate", beyond Galileo and Viviani, shortly afterwards involved Torricelli too, who dealt again with that argument in his **De Motu gravium naturaliter descendentium** (drawn up in 1640-41 but published in Florence in 1644 in the **Opera Geometrica**).[52] At the beginning of the first book, Torricelli starts with a recognition to Galileo: «Here we like to tackle (aggredi libet) the science of the motion of heavy bodies and projectiles, by many dealt with indeed, but, to the best of my knowledge, geometrically demonstrated only by Galileo. I admit he reaped all this harvest as by a sickle, and nothing remains to us except to follow the tracks of the industrious reaper and to pick up the left behind, or neglected ears, if any. At least we shall pick the privets and the humble violas, but maybe we shall entwine a non-contemptible garland of flowers».[53]

And next:

«On the point of dealing with the naturally accelerated motion Galileo assumes a principle, which he too doesn't consider fully evident, since he strives to prove it by means of

---

[52]**Opera Geometrica Evangelistae Torricelli**—Florentiae Typis Amatoris Massae & Laurentij de Landis 1644.

[53]Our translation.

the little accurate experiment of the pendulum, that is that the degrees of velocity of the same moveable acquired on differently inclined planes are equal, when the heights of the same planes are equal.

On this petition almost his doctrine both of the accelerated motion and of projectiles is depending.»

It is easy to deduce that Torricelli doesn't want to found the treatment of the motions on "little accurate" mechanical experiments; he rather wants to start from general assertions evident to everyone. He will say, a little further on:

«.... Galileo's supposition can be demonstrated, and immediately indeed, making use of that theorem he himself gathers as demonstrated from **Le Mechaniche** in the second part of the sixth Proposition of the accelerated motion, that is: "The momenta of equal heavy bodies on differently inclined planes are in the same ratio as the perpendiculars of equal parts of the same planes".»

At last,

«..... Since we did not meet with such a theorem, we shall confirm it through a demonstration, then we shall certainly pass to demonstrate what in Galileo is a principle or a petition.»

Hence Torricelli will construct his demonstration of the "postulate" by starting from a preliminary which extends Proposition I of the **Mechanicorum Liber** (1577) of Guidobaldo del Monte to a system composed of two heavy bodies[54]:

«Two heavy bodies joined together cannot move by themselves if their common centre of gravity doesn't fall.»

5.5 The Motion of Projectiles and the Parabolic Trajectory

We have seen that at last Galileo has achieved the correct law of fall of heavy bodies (the natural motion "deorsum") which he pursued since the time of **De Motu**.

For finishing the work it remained to find the law for what Aristotle called violent or forced motion. The latter had two protagonists: the projector (that is who, or which was the cause of motion) and the projectile (that is the moveable). In Aristotle's opinion the motion could happen only within a material medium, indeed, the medium sustained the motion.

On the contrary, Galileo soon abandoned the antiperistasis' theory and subsequently dealt with the motions as if they happened in vacuo, warning that one should leave out of consideration all the impediments due to the medium.

As often as he describes one of his experiments, he takes care to relate how he has minimized the friction (for instance, on an inclined plane).

---

[54]See **Le Mechaniche** dell'illustriss. Sig. Guido Ubaldo De' Marchesi del Monte—tradotte in volgare dal Signor Filippo Pigafetta—IN VENETIA, Appresso Evangelista Deuchino, MDCXV, pp. 6–7. An abridged translation of this work (by Stillman Drake) is in **Mechanics in Twentieth-Century Italy** edited by Stillman Drake & I.E. Drabkin—The University of Wisconsin Press, Madison, 1969.

In addition to what we can call an intellectual necessity of finding the correct explanation of the motion of a projectile, already before the time of Galileo's activity, the "practical" necessity of having precise rules for the trajectories of the cannon balls had arrived. We have already met with this argument when we were interested in the work of Niccolò Tartaglia. This is one of the so many cases in which the "war industry" asks the scientific research for help.

While for us today, basing ourselves on the Newtonian mechanics, the equation of the trajectory is obtained by starting from the time equations of uniform rectilinear motion and of the uniformly accelerated motion, for Galileo the way has been more complex.

As it is demonstrated in an essay of J. Renn, P. Damerow and S. Rieger,[55] the form of the trajectory has been established before the law of the uniformly accelerated motion (very likely in 1592). This date (1592) would prove, on the other hand, what Galileo maintains in the letter to Cesare Marsili of the eleventh of September 1632 («…. A study of mine of more than forty years …»).

1532 is the year in which one assumes Galileo visited Guidobaldo del Monte, in his residence of Mombaroccio (at that time Monte Baroccio), on route to Venice. In this occasion, Galileo and his sponsor should have performed a singular experiment, after all rather elementary, in order to describe the trajectory of a projectile.

A trace of this can be found in a Guidobaldo's manuscript that G. Libri discovered in the National Library of Paris.[56]

Let us see an excerpt of it for the argument we are interested in:

> « If one throws a ball with a catapult or with artillery or by hand or by some other instrument above the horizontal line, it will take the same path in falling as in rising, and the shape is that which, when inverted under the horizon, a rope makes which is not pulled, both being composed of the natural and the forced, and it is a line which in appearance is similar to a parabola and hyperbola …. The experiment of this movement can be made by taking a ball coloured with ink, and throwing it over a plane of a table which is almost perpendicular to the horizontal.
>
> Although the ball bounces along, yet it makes points as it goes, from which one can clearly see that as it rises so it descends, and it is reasonable this way, since the violence it has acquired in its ascent operates so that in falling it overcomes, in the same way, the natural movement in coming down.
>
> (op. cit. pp. 397–398)».[57]

---

[55]Jürgen Renn, Peter Damerow, Simone Rieger with an appendix by Domenico Giulini—**Hunting the white Elephant: When and How did Galileo Discover the Law of fall?** Special issue of Science in Context 13.2000 (pp. 29–149).

[56]It is a manuscript entitled **Meditatiunculae Guidi Ubaldi e marchionibus Montis Sanctae Mariae, de rebus mathematicis** then reported by Libri in the volume IV of his work **Histoire des Sciences Mathematiques …**, op. cit. pp. 369–398. In this regard see also: Martin Frank—**A proposal for a new dating of Guidobaldo del Monte's Meditatiunculae**—Bollettino di storia delle Scienze matematiche, XXXIII, 275-326, 2013.

[57]J. Renn et al., pp. 44–45.

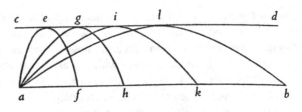

Fig. 5.11

As one can see, Guidobaldo is still hesitant between parabola and hyperbola and, moreover, he also equates the trajectory with an upturned catenary (a mistake shared by Galileo as well).

Only later Galileo will admit that the catenary is not equable to a parabola, when the two points where the chain is hanged are at a distance smaller than the vertical distance of the lowest point.

In the Paduan period, Galileo had frequent words also with Paolo Sarpi, as we have already remembered, both on the law of fall of heavy bodies, and the trajectory of projectiles. Paolo Sarpi left several thoughts on this argument and it is very interesting to compare these thoughts[58] with the contemporaneous thoughts of Galileo. It is also possible to find in them a hint to the composition of the two motions (uniform and uniformly accelerated), which will be the great novelty of Galileo.

A proof of the fact that Galileo always continued to attend to the motion of projectiles is the well-known letter to Antonio De' Medici of the eleventh of February 1609, where at a certain point he says:

«I am now working on some remaining questions about the motion of projectiles, among them many are relevant to the shots of artillery: and still recently I have found this fact, that putting the gun on an elevated place in open country, and pointing it on a level, the ball gone out from the gun, pushed either by a lot of powder or by a little or even by the only quantity necessary to push it out of the gun, comes always declining and bending downwards with the same velocity.

As a consequence, in the same time and for all flush shots, the ball arrives at ground in the case of both the longest and the shortest shots and even if the ball only goes out from the gun and suddenly falls in the plane of the country.

And the same thing occurs for the elevated shots, which are dispatched all in the same time, as long as they arrive at the same vertical height: like, for instance (Fig. 5.11), shots aef, agh, aik, alb, comprised between the same parallel lines cd, ab, are dispatched all in the same time; and the ball takes as much time in covering the line aef as in covering aik, and in covering any else; and as a consequence their halves, that is, parts ef, gh, ik, lb, are covered in equal times, which correspond to flush shots.»[59]

Even if in the text he does not explicitly indicate the parabolic form, the picture speaks for itself.

---

[58]Paolo Sarpi **Pensieri naturali, metafisici e Matematici**, a cura di L. Cozzi e L Sosio. Ricciardi, 1996—p. 398.

[59]E. N. X, pp. 229–230. Our translation.

Finally in the letter to Belisario Vinta (secretary of state of Ferdinand I) of the seventh of May 1610, on the point of coming back to Florence, while listing the works in progress he writes:

«I have plenty of particular secrets, both useful and of curiosity and admiration, and the consequent abundance damages and ever damaged me; the reason is that if I had known only one of them, I would have greatly appreciated it and with it I could have skipped forward, beside some great prince, and have met that fortune that till now I have neither met nor sought …. The works I have to carry trough are chiefly two books **De sistemate seu constitutione universi**, an immense concept full of philosophy, astronomy and geometry: three books **De motu locali**, a completely new science, since nobody else, ancient or modern, has discovered any of the great deal of admirable phenomena which I show to be present in the natural and violent motions, therefore I can, for a lot of reasons, call it a new science found by me starting with the fundamental principles: three books of mechanics, two pertaining to demonstrations and foundations, and one to problems; and even if others have written on the same matter, however, all what has been written till now is not a quarter, both for quality and quantity, of what written by me.»[60]

Yet twenty two years will have to pass before the blunder of Bonaventura Cavalieri, with his **Lo Specchio Ustorio**, make public (by the publication of the book) the fact that Galileo had "discovered" that the trajectory of projectiles is parabolic. Thinking about it, the waiting on the part of Galileo to publish the yielded result, even though it was known among his disciples as the letter of Cavalieri attests, appears to be exaggerated.

Most probably, he has waited until he has elaborated a suitable way to expose it. The awaited exposition will appear in the fourth Day of the **Discourses**:

«*On the Motion of Projectiles*

We have considered properties existing in equable motion, and those in naturally accelerated motion over inclined planes of whatever slope. In the studies on which I now enter, I shall try to present certain leading essentials, and to establish them by firm demonstrations, bearing on a moveable when its motion is compounded from two movements; that is, when it is moved equably and is also naturally accelerated. Of this kind appear to be those which we speak of as projections, the origin of which I lay down as follows.

I mentally conceive of some moveable projected on a horizontal plane, all impediments being put aside. Now it is evident from what has been said elsewhere at greater length that equable motion on this plane would be perpetual if the plane were of infinite extent; but if we assume it to be ended, and [situated] on high, the moveable (which I conceive of as being endowed with heaviness), driven to the end of this plane and going on further, adds on to its previous equable and indelible motion that downward tendency which it has from its own heaviness. Thus there emerges a certain motion, compounded from equable horizontal and from naturally accelerated downward [motion], which I call "projection." We shall demonstrate some of its properties [accidentia], of which the first is this:

PROPOSITION I. THEOREM I

When a projectile is carried in motion compounded from equable horizontal and from naturally accelerated downward [motions], it describes a semiparabolic line in its movement.»[61]

---

[60]E. N. X, pp. 351–352. Our translation.

[61]E. N. VIII, pp. 268–269. (Drake p. 217).

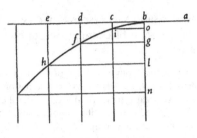

Fig. 5.12

Futher on, Salviati says:

«SALV. So let us take up this text again, and see how he demonstrates his first proposition, in which his purpose is to prove to us that:

[THEOREM I, restated]

The line described by a heavy moveable, when it descends with a motion compounded from equable horizontal and natural falling [motion], is a semiparabola.

Imagine a horizontal line or plane ab (Fig. 5.12) situated on high, upon which the moveable is carried from a to b in equable motion, but at b lacks support from the plane, whereupon there supervenes in the same moveable, from its own heaviness, a natural motion downward along the vertical bn. Beyond the plane ab imagine the line be, lying straight on, as if it were the flow or measure of time, on which there are noted any equal parts of time bc, cd, be; and from points b, c, d, and e imagine lines drawn parallel to the vertical bn. In the first of these, take some part ci; in the next, its quadruple df; then its nonuple eh, and so on for the rest according to the rule of squares of cb, db, and eb; or let us say, in the duplicate ratio of those lines.

If now to the moveable in equable movement beyond b toward c, we imagine to be added a motion of vertical descent according to the quantity ci, the moveable will be found after time bc to be situated at the point i. Proceeding onwards, after time db (that is, double bc,), the distance of descent will be quadruple the first distance, ci; for it was demonstrated in the earlier treatise that the spaces run through by heavy things in naturally accelerated motion are in the squared ratio of the times.

And likewise the next space, eh, run through in time be, will be as nine times [ci]; so that it manifestly appears that spaces eh, df, and ci are to one another as the squares of lines eb, db, and cb. Now, from points i, f, and h, draw straight lines io, fg, and hl parallel to eb, line by line, hl, fg, and io will be equal to eb, db, and cb respectively, and bo, bg, and bl will be equal to ci, df, and eh. And the square of hl will be to the square of fg as line le is to bg, while the square of fg [will be] to the square of io as gb is to bo; therefore points i, f, and h lie in one and the same parabolic line.

And it is similarly demonstrated, assuming any equal parts of time, of any size whatever, that the places of moveables carried in like compound motion will be found at those times in the same parabolic line. Therefore the proposition is evident.

SALV. This conclusion is deduced from the converse of the first of the two lemmas given above. For if the parabola is described through points b and h, for example, and if either of the two [points], f or i, were not in the parabolic line described, then it would lie either inside or outside, and consequently line fg would be either less or greater than that which would go to terminate in the parabolic line. Whence the ratio that line lb has to bg, the square of hl would have, not, to the square of fg, but to [the square of] some other [line] greater or less [than fg]. But it [the square of hl] does have [that ratio] to the square of fg. Therefore point f is on the parabola; and so on for all the others, etc.

SAGR. It cannot be denied that the reasoning is novel, ingenious, and conclusive, being argued ex supposition it is, by assuming that the transverse motion equable, and that the natural downward [motion] likewise maintains its tenor of always accelerating according to the squared ratio of the times; also that such motions, or their speeds, in mixing together do not alter, disturb, or impede one another. In this way, the line of the projectile, continuing its motion, will not finally degenerate into some other kind [of curve]. But this seems to me impossible; for the axis of our parabola is vertical, just as we assume the natural motion of heavy bodies to be, and it goes to end at the center of the earth. Yet the parabolic line goes ever widening from its axis, so that no projectile would ever end at the center [of the earth], or if it did, as it seems it must, then the path of the projectile would become transformed into some other line, quite different from the parabolic.»[62]

As Salviati states precisely, the conclusion on the part of Galileo, that the curve is a parabola, comes from the two propositions he has previously demonstrated (not reported here). The treatise on the conics of Apollonius, together with Euclid's **Elements**, was considered at that time a text that a mathematician should absolutely know and Galileo (as, later on, Newton) was able to apply it to deduce the two propositions relevant to his problem.

A posteriori one is even tempted to justify the reluctance of Galileo to make his ideas on the trajectory of projectiles official. In fact, many people questioned the fact that, in the composition, the two motions (both the uniform rectilinear motion and the uniformly accelerated motion) remained unchanged. For some people, the rectilinear motion out of the plane could not maintain the same velocity it had on the plane, independently from the gravity. In a similar way, they believed that the accelerated motion of fall should occur at a smaller velocity.

The more destructive criticism, anyway, came from Descartes, who wrote a letter to Mersenne (a letter of the October 1638, well known in literature) in which, after having given a negative judgement on several fundamental points of the **Discourses**, he writes:

«For the rest, please be good enough to hold these remarks only for you, who wanted me to write them, you, towards whom I have so many obligations that I think I cannot refuse you anything in my power.»[63]

After the Proposition I we have reported, Galileo deals with the problem in full until reaches the result, already found by Tartaglia, that

«Hence it follows that if projections are made with the same impetus from point D, but according to different elevations, the maximum projection, or amplitude of semiparabola

[62]E. N. VIII, pp. 272–274; Drake, pp. 221–223.

[63]René Descartes—**Oeuvres de Descartes**—Correspondence—op. cit.

(or whole parabola) will be that corresponding to the elevation of half a right angle. The others, made according to larger or smaller angles will be shorter.»[64]

Even the corollary (which follows Proposition VII), wherefrom the passage quoted above has been extracted, has not been except from Descartes' criticism.[65]

According to many people, Torricelli become acquainted with it, maybe through Mersenne; then he tried to demonstrate what had been questioned in the second book of his Opera Geometrica (1644).[66]

The last part of the fourth Day of the **Discourses** is dedicated to the problems relevant to the shots of artillery and includes also tables. These problems had already been mentioned, with examples, also in the **Dialogue**, since they were undoubtedly always present to Galileo given his role of Mathematician of the Grand Duke.

What we have tried to point out is that, notwithstanding the long gestation of the problem of the motion of projectiles, at last Galileo has derived the parabolic form of the trajectory by composing the two motions (uniform rectilinear and uniformly accelerated) and resorting to a geometric demonstration. The composition of motions was already appeared in the second Day of the **Dialogue** (a body falling from a tower on the Earth during the diurnal rotation) but only here he uses "mathematically" the "equations of motion" to obtain the curve of the trajectory.

5.6 A Gloss on the Pendulum

At this point, the reader who has followed us up to here feels urged to ask us: and the pendulum?

It's true, to tell the truth, so far we have not dealt with it. The reason is that, even if the pendulum (or better, the use of the traditional schematization of the simple pendulum) appears many times in Galileo's work, in our opinion it does not hold the important function one has attached to it with regard to the discovery of the law of fall of the heavy bodies. Nevertheless, we don't want to evade a digression on the argument.

Unfortunately, when one begins to speak about the pendulum in relation to Galileo, he finds himself to reckon with an anecdotic literature built up over the centuries starting from the **Historical Account of the Life of Galileo Galilei** by Viviani, where the "historical" account often digresses in the legend.

At the beginning there is the well-known passage:

«In the meantime with the sagacity of his mind invented that simplest and regulated measure of the time by means of the pendulum, by nobody else invented before, having had the opportunity to deduce it from the observation of the motion of a lamp, while he was a day

---

[64]E. N. VIII, p. 296. Drake, p. 245.

[65]René Descartes, loc. cit.

[66]**Opera Geometrica** ... op. cit. pp. 157–158.

in the Cathedral of Pisa and, by doing very accurate experiments, checked the equality of its vibrations, and then the idea came to his mind of adapting it to the practice of medicine for the measure of the frequency of the pulse, with astonishment and pleasure from the physicians of those times and how also today it is popularly practised: then he made use of this invention in several experiments and measures of times and motions, and was the first to apply it to celestial observations, with astounding advantage in the astronomy and geometry.»[67]

The same thing was practically repeated in the "Eulogy" of Galileo by Paolo Frisi.[68]

The fact that the lamp, that the guides indicate to tourists as "Galileo's lamp", has been installed four years after Galileo left the University of Pisa, and therefore has the same historical veridicalness of "Juliet's balcony" in Verona, has not prevented people from debating over the centuries the invention of the pendulum-clock on the part of Galileo.[69]

It must be said, anyway, that similar news were spreading over in Europe already before the **Historical Account** of Viviani was printed. In fact, Viviani's book came out (delayed) only in 1717 and next reprinted in the edition of Galileo's works of 1718. Maybe it is unfair to hold him responsible of a legend of which he had not been the only author.

Nevertheless, one can say that Viviani had legitimated it.

It is known that already Huygens had argued with those people who considered Galileo a designer of pendulum-clocks. He said in fact, in the preface of his **Horologium Oscillatorium** (1673):

«Qui vero Galileo primus hic deferre conantur, si tentasse eum, non vero perfecisse inventum dicavit, illius magis quam meae laudi detrahere videntur, quippe qui rem eandem, meliore quam ille eventu, investigarerim.»[70]

Once blown up the echoes of the discussions about the invention of the clock, in more recent times they were replaced by the discussion about the experiments regarding the isochronism of the oscillations, which Galileo deals with on several occasions starting from the well-known letter to Guidobaldo del Monte of November twenty-nine 1602, where he persists in trying to persuade him of his results. Let us see a passage of it:

«….. Then I take two thin threads (Fig. 5.13), equally long two or three braccia each, and let they be AB, EF, and I hang them up from two nails A, E, and in the other ends I tie two

---

[67]Vincenzio Viviani: **Vita di Galileo**, op. cit. pp. 32–33. Our translation.

[68]Paolo Frisi: **Elogio del Galileo**, in Livorno, MDCCLXXV, p. 14.

[69]See, for example, in the supplement to the already quoted edition of Galileo's works edited by E. Albèri (1856), pp. 331–358, "on the pendulum-clock of Galileo Galilei and on two recent divinations of the mechanism he imagined".

[70]**Christiani Hugenii: Orologium Oscillatorium** …Parisiis, MDCCXXIII, p. 3. «If those who desire to refer the origins of this matter to Galileo say that he tried but failed to make this invention, it seems that they detract praise from him rather than from me, since indeed in that case I would have carried out this same investigation more successfully than he did.» Translated with notes by R.J. Blackwell—Iowa Stale University Press, 1986, p. 13.

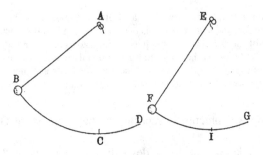

Fig. 5.13

equal lead balls (it would be the same even if they were different); then, removing each of these threads from his perpendicular, but one very much along the arc GB, and the other very little, along the arc IF, I let them go freely simultaneously, and the one begins to make great arcs, like the arc BCD, and the other makes little arcs, like in figure; but the moveable B (for covering all the arc BCD) does not take more time than the other moveable to cover the arc in figure make me most sure of this in this way: ....»[71]

Experiments involving pendulums are described both in the **Dialogue** (E. N. VII, pp. 256–257, 474–476) and in the **Discourses** (E. N. VIII, pp. 128–129, 139–140, 277–278).

Before we report revealing passages of the works, let us try to understand why these experiments have been so much discussed and challenged in the last decades.

Today, we know that the oscillations of the simple pendulum (the present definition—a mass point hanging from an inextensible and weightless thread—agrees with Galileo's descriptions) are ruled by a nonlinear differential equation of the second order (obviously by neglecting the air resistance). Exactly, if we call l the length of the thread and ϑ the angle that the thread makes with the perpendicular during the oscillations, the equation is $\frac{d^2\vartheta}{dt^2} + \frac{g}{l}\sin\vartheta = 0$, where g is the acceleration of gravity.

Unfortunately, since we have to do with a nonlinear differential equation, the solution cannot be expressed by means of simple functions. As a consequence, one can obtain the period of the oscillations only through a complete elliptic integral K. One obtains in fact, for the period T: $T = 4\sqrt{\frac{l}{g}}\int_0^{\pi/2}\frac{d\vartheta}{1-k^2\sin^2\vartheta}$ which, when the modulus $k = \sin\frac{\vartheta}{2}$ tends to zero, becomes $T = 2\pi\sqrt{\frac{l}{g}}$.

The last expression corresponds to the equation $\frac{d^2\vartheta}{dt^2} + \frac{g}{l}\vartheta = 0$, which is the equation of the so called "linearized pendulum", that is with $\sin\vartheta$ replaced by the first term (ϑ) of its series expansion.

Therefore, for the small oscillations, that is, for $\sin\vartheta \cong \vartheta$, one can say with a good approximation that the period of oscillations is given by $T = 2\pi\sqrt{\frac{l}{g}}$ and then depends only on the length l of the pendulum and not on the amplitude ϑ.

---

[71]E. N. X, p. 98 (our translation).

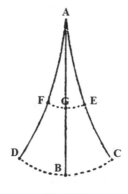

Fig. 5.14

The problem arises because, in the opinion of Galileo, this was valid (always with a good approximation) for great ϑ as well (even 80°).

In fact, in the two passages of the **Dialogue** we have mentioned above, he says:

«SALV. Tell me: of two pendulums of unequal length, doesn't the one which is hanging by the longer cord perform its oscillations the more infrequently?

SAGR. Yes, if they are swinging an equal distance from the perpendicular.

SALV. Oh, that makes no difference, for the same pendulum makes it oscillations in equal times, whether they are long or short (that is, whether the pendulum is removed a long way or very little from the perpendicular).

Or, if they are not exactly equal, the difference is insensible, as experiment will show you. But even if they were quite unequal, that would help rather than hinder my case.

For let us denote the perpendicular AB (Fig. 5.14), and hang from the point A on the cord AC the weight C, and still another, higher up on the same, which shall be E. Drawing the cord AC aside from the perpendicular and letting it loose, the weights C and E will move through the arcs CBD and EGF, and the weight E, hanging at the lesser distance and also being moved aside less, as you said, would try to go back sooner and to make its vibrations more frequently than the weight C. Therefore it would impede the latter from going back as far toward the point D as it would do if it were free, and, being thus an impediment to it in every oscillation, would finally bring it to rest.

Now this cord, with the middle weight removed, is itself a compound of many weighted pendulums; that is, each of its parts is just such a pendulum, attached closer and closer to the point A, and therefore arranged so as to make its vibrations more and more frequent, and consequently each is able to place a continual hindrance on the weight C. An indication of this is that as we observe the cord AG, we see it stretch not tightly, but in an arc; and if in place of the cord we put a chain, we see this effect much more evidently; most of all when the weight C is quite far from the perpendicular AB. For the chain is composed of many linked parts, each of which is heavy, and the arcs AEG and AFD will be seen to be noticeably curved. Therefore since the parts of the chain try to make their vibrations the more frequent according to their closeness to the point A, the lowest part cannot travel as much as it would naturally. And with the continual lessening of the vibrations of the weight C, they would finally stop even if the impediment of the air were taken away.»[72]

---

[72]E. N. VII, pp. 256–257; Drake, pp. 267–268.

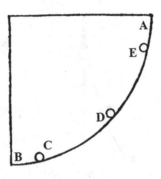

Fig. 5.15

And again in the fourth Day:

«The other particular is truly remarkable; it is that the same pendulum makes its oscillations with the same frequency, or very little different almost imperceptibly whether these are made through large arcs or very small ones along a given circumference.

I mean that if we remove the pendulum from the perpendicular just one, two, or three degrees, or on the other hand seventy degrees or eighty degrees, or even up to a whole quadrant, it will make its vibrations when it is set free with the same frequency in either case; in the first, where it must move only through an arc of four or six degrees, and in the second where it must pass through an arc of one hundred sixty degrees or more.

This is seen more plainly by suspending two equal weights from two threads of equal length, and then removing one just a small distance from the perpendicular and the other one a very long way. Both, when set at liberty, will go back and forth in the same times, one by small arcs and the other by very large ones.

From this follows the solution of a very beautiful problem, which is this: Given a quarter of a circle (Fig. 5.15) I shall draw it here in a little diagram on the ground which shall be AB here, vertical to the horizon so that it extends in the plane touching at the point B; take an arc made of a very smooth and polished concave hoop bending along the curvature of the circumference ADB, so that a well-rounded and smooth ball can run freely in it (the rim of a sieve is well suited for this experiment).

Now I say that wherever you place the ball, whether near to or far from the ultimate limit B placing it at the point C, or at D, or at E and let it go, it will arrive at the point B in equal times (or insensibly different), whether it leaves from C or D or E or from any other point you like; a truly remarkable phenomenon.

Now add another, no less beautiful than the last. This is that along all chords drawn from the point B to points C, D, E, or any other point (taken not only in the quadrant BA, but in the whole circumference of the entire circle), the same movable body will descend in absolutely equal times. Thus in the same time which it takes to descend along the whole diameter erected perpendicular to the point B, it will also descend along the chord BC even when that subtends but a single degree, or yet a smaller arc.»[73]

Where we have limited the quotation of Salviati's talk to the part where there is the assertion that the period of the pendulum depends entirely on the length even in case of oscillations close to 80°.

---

[73]E. N. VII, pp. 475–476; Drake, pp. 522–523.

The assertion is reiterated many times in the **Discourses**.
Let us start from the first Day:

«SALV. ….. ultimately, I look two balls, one of lead and one of cork, the former being at least a hundred times as heavy as the latter, and I attached them to equal thin strings four or five braccia long, tied high above. Removed from the vertical, these were set going at the same moment, and falling along the circumferences of the circles described by the equal strings that were the radii, they passed the vertical and returned by the same path.

Repeating their goings and comings a good hundred times by themselves, they sensibly showed that the heavy one kept time with the light one so well that not in a hundred oscillations, nor in a thousand, does it get ahead in time even by a moment, but the two travel with equal pace.

The operation of the medium is also perceived; offering some impediment to the motion, it diminishes the oscillations of the cork much more than those of the lead. But it does not make them more frequent, or less so; indeed, when the arcs passed by the cork were not more than five or six degrees, and those of the lead were fifty or sixty, they were passed over in the same times.

SIMP. If that is so, why then will the speed of the lead not be [called] greater than that of the cork, seeing that it travels sixty degrees in the time that the cork hardly passes six?

SALV. And what would you say, Simplicio, if both took the same time in their travels when the cork, removed thirty degrees from the vertical, had to pass an arc of sixty, and the lead, drawn but two degrees from the same point, ran through an arc four?

Would not the cork then be as much the faster? Yet experience shows this to happen. But note that if the lead pendulum is drawn, say, fifty degrees from the vertical and released, it passes beyond the vertical and runs almost another fifty, describing an are of nearly one hundred degrees. Returning of itself, it describes another slightly smaller are; and continuing its oscillations, after a great number of these it is finally reduced to rest. Each of those vibrations is made in equal times, as well that of ninety degrees as that of fifty, or twenty, or ten, or of four. Consequently the speed of the moveable is always languishing, since in equal times it passes successively arcs ever smaller and smaller.

A similar effect, indeed the same, is produced by the cork that hangs from another thread of equal length, except that this comes to rest in a smaller number of oscillations, as less suited by reason of its lightness to overcome the impediment of the air.

Nevertheless, all its vibrations, large and small, are made in times equal among themselves, and also equal to the times of the vibrations of the lead.

Whence it is true that if, while the lead passes over an arc of fifty degrees, the cork passes over only ten, then the cork is slower than the lead; but it also happens in reverse that the cork passes along the arc of fifty while the lead passes that of ten, or six; and thus, at different times, the lead will now be faster, and again the cork. But if the same moveables also pass equal arcs in the same equal times, surely one may say that their speeds are equal.»[74]

Further on, he will specify the relation which associates the period of the pendulum to its length:

---

[74]E. N. VIII, pp. 128–130; Drake, pp. 87–88.

«As to the ratio of times of oscillation of bodies hanging from strings of different lengths, those times are as the square roots of the string lengths; or we should say that the lengths are as the doubled ratios, or squares, of the times. Thus if, for example, you want the time of oscillation of one pendulum to be double the time of another, the length of its string must be four times that of the other; or if in the time of one vibration of the first, another is to make three, then the string of the first will be nine times as long as that of the other. It follows from this that the lengths of the strings have to one another the [inverse] ratio of the squares of the numbers of vibrations made in a given time.»[75]

At last, in the fourth Day he reaffirms the equality of the periods with a long description of the experiment:

«SALV. ... Suspend two equal lead balls from two equal threads four or five braccia long. The threads being attached above, remove both balls from the vertical, one of them by 80 degrees or more, and the other by no more than four or five degrees, and set them free. The former descends, and passing the vertical describes very large [total] arcs of 160°, 150°, 140°, etc., which gradually diminish. The other, swinging freely, passes through small arcs of 10°, 8°, 6°, etc., these also diminishing bit by bit.

I say, first, that in the time that the one passes its 180°, 160°, etc., the other will pass its 10°, 8°, etc. From this it is evident that the [overall] speed of the first ball will be 16 or 18 times as great as the [overall] speed of the second; and if the greater speed were to be impeded by the air more than the lesser, the oscillations in arcs of 180°, 160°, etc. should be less frequent than those in the small arcs of 10°, 8°, 4°, and even 2° or one degree.

But experiment contradicts this, for if two friends shall set themselves to count the oscillations, one counting the wide ones and the other the narrow ones, they will see that they may count not just tens, but even hundreds, without disagreeing by even one; or rather, by one single count.

This observation assures us of both [the above] propositions at once; that is, that the greatest and least oscillations all are made, swing by swing, in equal times, and that the impediment and retardation of the air does no more in the swiftest [of these] motions than in the slowest, contrary to what all of us previously believed.»[76]

A few years after the publication of the **Discourses**, some discussions on the experiments made by Galileo started, particularly in France; they involved Mersenne, Gassendi and some Jesuits and Roberval as well. Then, the discussion continued in Italy, involving Baliani and Torricelli.

We mention this circumstance since, in the criticisms to the experiments described by Galileo, also the objections regarding the isochronism of the pendulum oscillations were comprised.[77]

Neglecting the criticisms typical of that time, which in part can also be due to an aptitude for experimentation lower than that of the more expert Galileo, let us come instead to the discussions of our days.

---

[75]E. N. VIII, pp. 139–140; Drake, pp. 97–98.

[76]E. N. VIII, pp. 277–278; Drake, pp. 226–227.

[77]See, for example, chap. XIX (De variis difficultatibus ad funependulum et casum gravium pertinentibus) of the work of Mersenne **Novarum observationum physico-mathematicorum— Tomus III**, pp. 152–159 (Parisiis, MDCXLVII) where he maintains "Funependuli vibrationes non esse isochronas".

We can start with Alexandre Koyré[78] and, after a goodly number of continuators (see the bibliography), stop at a well-known paper of Ronald Naylor,[79] who repeated the experiment of Galileo on pendulums.

Obviously, Naylor found that none of the pendulums he examined was perfectly isochronous but, in addition, he disputed the conclusions reached by Galileo on the different behaviours of light and weighty pendulums.

It was doubtful if Galileo reported precise results of experiments really performed in the form he told or, instead, reconstructed the experiments deducing from them not reproducible results.

Recently a "defence" of Galileo (experimenter) has been put forward by Paolo Palmieri, an historian of science of the last generation.

He started with a book **Renacting Galileo's Experiments**[80] in which, besides discussing Galileo's experiments, he has reconstructed their "experimental philosophy", and then dedicated a long paper to pendulum experiments.[81]

The aim of the paper is that of studying the discrepancy between an ideal pendulum (that is, a linearized one) and a real pendulum, to realize how Galileo could arrive at his conclusions. Besides repeating directly the experiments, trying to disclose their difficulties, Palmieri makes use of computer simulations and of the techniques of the nonlinear analysis as well.

He has also done a re-reading of the texts and pointed out (just referring to the Italian text) that Galileo never speaks of isochronism (term used—as we have seen—by Mersenne too), but says that the pendulums do their "reciprocations" always in equal times ("sotto tempi eguali").

Finally, he underlines that Galileo works on long times: that is, he compares the number of oscillations of pendulums (of the same length and different amplitude) occurred in the same (long) time. It's a matter of numbers of the order of a hundred and in case they can at most be imperceptibly different ("insensibilmente differenti").

This explains Galileo's conclusions.

With this we conclude this gloss, reaffirming that, in our opinion, the experiments with pendulums have not played an important role in the "discovery" of the law of fall of the heavy bodies and then in the uniformly accelerated motion.

We feel like subscribing the conclusion reached by Renn, Damerow and Rieger in the already quoted essay: «The law of fall, according to our reconstruction, was merely a trivial consequence of the recognition of the parabolic shape of the trajectory.»[82]

---

[78]A. Koyré: **An experiment in measurement**—Proceedings of the American Philosophical Society—97, 222–237, 1953.

[79]R. Naylor: **Galileo's simple pendulum**—Physis—16, 23–46, 1974.

[80]P. Palmieri: **Reenacting Galileo's Experiments: Rediscovering the Techniques of Seventh-Century Science**—The Mellen Press, 2008.

[81]P. Palmieri: **A Phenomenology of Galileo's experiments with Pendulum**—BJHS 42(4): 479–513, 2009.

[82]J. Renn, P. Damerow, S. Rieger—loc. cit. p. 30.

Bibliography

We quote here only books and papers not mentioned in the footnotes of the chapter.

P. Ariotti, *Galileo on the isochrony of the pendulum.* Isis **59**, 414–426 (1968)

S. Drake, *Free fall from Albert of Saxony to Honoré Fabri.* Stud. Hist. Phil. Sci. **5**, 347–366 (1976)

S. Drake, *History of Free Fall—Aristotle to Galileo. Published in combined edition with two new sciences* (Wall & Emerson, Inc., Toronto, 2008²)

S. Ducheyne, *Galileo and Huygens on free fall: mathematical and methodological differences.* Dinamis **28**, 243–274 (2008)

H. Erlichson, *Galileo's pendulums and planes.* Annals Sci. **51**, 263–272 (1994)

G. Galilei, *The Two Italian Editions of the Discourses: Discorsi e Dimostrazioni Matematiche Intorno a Due Nuove Scienze* a cura di Adriano Carugo e Ludovico Geymonat (Boringhieri, Torino, 1958)

Discorsi ... a cura di Enrico Giusti (Einaudi, Torino, 1990)

Paolo Galluzzi, *Gassendi and l'affaire Galilée of the laws of motion.* Sci. Context **14**, 239–275 (2001)

K. David Hill, *Galileo's early experiments on projectile motion and the law of fall.* Isis **79**, 646–668 (1988)

R. Naylor, *The evolution of an experiment: Guidobaldo Del Monte and Galileo's Discorsi demonstration of the parabolic trajectory.* Physis **16**, 323–346 (1974)

R. Naylor, *An aspect of Galileo's study of the parabolic trajectory.* Isis **66**, 394–396 (1975)

R. Naylor, *Galileo: the Search for the parabolic trajectory.* Annals Sci. **33**, 153–172 (1976)

R. Naylor, *Galileo: real experiment and didactic demonstration.* Isis **67**, 398–419 (1976)

R. Naylor, *Galileo's theory of projectile motion.* Isis **71**, 550–570 (1980)

R. Naylor, *Galileo's early experiments on projectile trajectories.* Annals Sci. **40**, 391–396 (1983)

M. Segre, *The role of experiment in Galileo's physics.* Arch. Hist. Exact Sci. **23**, 227–252 (1980)

C. Vilain, *La Loi Galiléenne et la Dinamique de Huygens.* Revue d'histoire des mathématiques **2**, 95–117 (1996)

Chapter 6
Galileo and the Principle of Relativity

Nowadays, in any textbook of physics (in the part devoted to mechanics) one finds written that the laws of Newtonian mechanics are invariant under Galileo transformations, or, equivalently, that the equations of the mechanics are the same in all inertial reference systems. As inertial system one think of a reference system (for instance represented by a tern of orthogonal Cartesian axes with the addition of a clock) where the inertia principle holds rigorously. Also, it is asserted that, if an inertial system exists, then an infinity of inertial system exist: all the possible systems which move with uniform rectilinear motion with respect to the given one.

The "Galileo transformations" are those which allow one to pass from an inertial system to another. If we have a system O, where the spatial coordinates are x, y, z and the time coordinate t, a system O' in motion with respect to O with constant velocity v in the direction of x-axis will have $x' = x \mp vt$, $y' = y, z' = z, t' = t$ (for the sake of simplicity, we have chosen the case in which the relative motion is along the x-axis without specifying the sense).

The laws of motion expressed in the coordinates x', y', z', t' will be the same of those expressed in the preceding coordinates, since the equations of motion are differential equations of the second order and the insertion of a constant velocity (therefore with vanishing time derivative) does not change anything. This is, in synthesis, the principle of Galilean relativity.

It is clear that Galileo has not expressed it in these terms. In the present formalism, the invariance under the Galilean transformations is immediately verified; we could even say that it is obvious.

We shall see, instead, how it can emerge from Galileo's writing. To do this, we shall have to see how the relative motion was dealt with before Galileo.

© Springer International Publishing Switzerland 2016
D. Boccaletti, *Galileo and the Equations of Motion*,
DOI 10.1007/978-3-319-20134-4_6

6.1 The Relative Motion Before Galileo

In the modern mechanics, we are used to study the motions with respect to different reference systems and to make use of the relations which make one pass from a system to another one. In the ancient times this problem did not appear and the motion of any body was considered exclusively with respect to the observer, who implicitly was the reference system. We can say that the only exception to this was represented by the case in which one agreed with the heliocentric hypothesis of Aristarchus of Samos (III cent. B.C.).

In this case, in fact, one had to face the problem of pointing out the motion of the Earth (no more the centre of the world) with respect to the Sun.

This problem was particularly discussed in the Middle Ages starting from the XIV century. We find a trace of this, for instance, in the work of Buridan, in which it is discussed (with reference to **On the Heavens** of Aristotle) "If the Earth is always still at the centre of the world" (question XXII).

In attributing to the motion of the Earth, rather than to that of the Sun, the generation of the night and of the day, Buridan justifies the fact that we do not notice it in this way:

«Et potestis de hoc accipere exemplum; quia si aliquis movetur in navi et imaginetur se quiescere, et videat aliam navem quae secundum veritatem quiescit, apparebit sibi quod illa alia navis moveatur; quia omnino taliter se habebit oculus ad illam aliam navem, si propria navis quiescat et alia moveatur, sicut se haberet si fieret e contrario. Et ita etiam ponamus quod sphaera solis omnino quiescat, et terra portando nos circumgiretur; cum tamen imaginemur nos quiescere, sicut homo existens in navi velociter mota non percipit motum suum nec motum navis, tunc certum est quod ita sol nobis oriretur et postea nobis occideret sicut modo facit quando ipse movetur et nos quiescimus.».[1]

Obviously, Buridan proceeds with extreme caution in exposing the considerations in favour of the heliocentric hypothesis. The time is not yet ripe to doubt the geocentric model. We do not intend to go into detail of the exposition of the different examples which were adduced against the diurnal motion of the Earth, since our purpose is only of seeing how the problem of the relative motion was tackled.

After Buridan, Oresme (in 1477), in dealing with the same problem, is some more didactic:

«For example, if a man is in a boat a, which is moving very smoothly either at rapid or slow speed, and if this man sees nothing except another boat b, which moves precisely like boat a, the one in which he is standing, I maintain that to this man it will appear that

---

[1]See: Johannis Buridani: **Quaestiones super libros quattuor de caelo et mundo**—ed. E. A. Mody, Cambridge (Mass.), 1942. «And you could have an example of this; if one moves in a ship and would imagine of being still, and see another ship really still, that other will appear to him as moving; since the age will completely refer to that other ship, if his own ship is still and the other moves, in the same way as he would do if the contrary would occur. And so let us suppose that the sphere of the Sun is completely still and the Earth circling around it; if we imagine to be still, like a man on a ship which moves speedily does not perceives his motion neither the motion of the ship, the nit is certain that in that case for us the Sun before would rise and after would set in the same way as when it moves and we are still.» (Our translation).

neither boat is moving. If *a* rests while *b* moves, he will be aware that *b* is moving; if *a* moves and *b* rests, it will seem to the man in *a* that *a* is resting and *b* is moving, just as before. Thus, if *a* rested an hour and *b* moved, and during the next hour it happened conversely that *a* moved and *b* rested, this man would not be able to sense this change or variation; it would seem to him that all this time *b* was moving. This fact is evident from experience, and the reason is that the two bodies *a* and *b* have a continual relationship to each other so that, when *a* moves, *b* rests and, conversely, when *b* moves, *a* rests.»[2]

At last, almost a century later, Nicholas of Cusa, in his most known work:

«For example, if someone did not know that a body of water was flowing and did not see the shore while he was on a ship in the middle of the water, how would he recognize that the ship was being moved? And because of the fact that it would always seem to each person (whether he were on the earth, the sun, or another star) that he was at the "immovable" center, so to speak, and that all other things were moved: assuredly, it would always be the case that if he were on the sun, he would fix a set of poles in relation to himself; if on the earth, another set; on the moon, another; on Mars, another; and so on.»[3]

Also this last passage is in a chapter entitled "De conditionibus terrae" i.e. it is always the diurnal motion of the Earth to be discussed. And so it will be for Galileo too, in the second Day of the **Dialogue**.

6.2 How Galileo Expresses the Principle of Relativity

Of course, Galileo too, as his predecessors, resorts to the example of a ship in motion with respect to the shore, most probably because it was the most suitable example, at that time, for suggesting a uniform rectilinear motion.

We quote the most significant part of that «Experiment which alone shows the nullity of all those adduced against the motion of the Earth»:

«SALV. ..... For a final indication of the nullity of the experiments brought forth, this seems to me the place to show you a way to test them all very easily. Shut yourself up with some friend in the main cabin below decks on some large ship, and have with you there some flies, butterflies, and other small flying animals. Have a large bowl of water with some fish in it; hang up a bottle that empties drop by drop into a narrow-mouthed vessel beneath it. With the ship standing still, observe carefully how the little animals fly with equal speed to all sides of the cabin. The fish swim indifferently in all directions; the drops fall into the vessel beneath; and, in throwing something to your friend, you need throw 'it no more strongly in one direction than another, the distances being equal; jumping with your feet together, you pass equal spaces in every direction. When you have observed all these things carefully (though there is no doubt that when the ship is standing still everything must happen in this way), have the ship proceed with any speed you like, so long as the motion is uniform and not fluctuating this way and that. You will discover not the least change in all the effects named, nor could you tell from any of them whether

---

[2]Oresme: **Le livre du ciel et du monde**—edited by A. D. Menut and A. J. Denomy—The University of Wisconsin Press, 1968, p. 523.

[3]Nicolas of Cusa: **De docta ignorantia**, II, 12 (Ed. Paris. fol 21v). (translation of Jasper Hopkins from Cusan-Portal).

the ship was moving or standing still. In jumping, you will pass on the floor the same spaces as before, nor will you make larger jumps toward the stern than toward the prow even though the ship is moving quite rapidly, despite the fact that during the time that you are in the air the floor under you will be going in a direction opposite to your jump. In throwing something to your companion, you will need no more force to get it to him whether he is in the direction of the bow or the stern, with yourself situated opposite. The droplets will fall as before into the vessel beneath without dropping toward the stern, although while the drops are in the air the ship runs many spans. The fish in their water will swim toward the front of their bowl with no more effort than toward the back, and will go with equal ease to bait placed anywhere around the edges of the bowl. Finally the butterflies and flies will continue their flights indifferently toward every side, nor will it ever happen that they are concentrated toward the stern, as if tired put from keeping up with the course of the ship, from which they will have been separated during long intervals by keeping themselves in the air. And if smoke is made by burning some incense, it will be seen going up in the form of a little cloud, remaining still and moving no more toward one side than the other. The cause of all these correspondences of effects is the fact that the ship's motion is common to all the things contained in it, and to the air also. That is why I said you should be below decks; for if this took place above in the open air, which would not follow the course of the ship, more or less noticeable differences would be seen in some of the effects noted. No doubt the smoke would fall as much behind as the air itself. The flies likewise, and the butterflies, held back by the air, would be unable to follow the ship's motion if they were separated from it by a perceptible distance. But keeping themselves near it, they would follow it without effort or hindrance; for the ship, being an unbroken structure, carries with it a part of the nearby air. For a similar reason we sometimes, when riding horseback, see persistent flies and horseflies following our horses, flying now to one part of their bodies and now to another. But the difference would be small as regards the falling drops, and as to the jumping and the throwing it would be quite imperceptible.»[4]

It is, as it is known, a fine prose passage, but we shall limit ourselves to isolate the scientific term. Galileo says: «…. so long as the motion is uniform and not fluctuating this way and that. You will discover not the least change in all the effects named, nor could you tell from any of them whether the ship was moving or standing still.»

That is, in the actual language, no mechanical experiment performed inside an inertial system can disclose its state of motion or of rest.

Through a more abstract formulation, the Principle will be confirmed, about half a century later, by Newton into his Corollary V: «Corporum dato spatio inclusorum iidem sunt motus inter se, sive spatium illud quiescat, sive moveatur idem uniformiter in directum sine motu circulari.»[5]

After Galileo, and before Newton, the principle of relativity was used by Huygens who, in his collision theory written in opposition to that of Descartes, maintained that the collision between two equal spheres occurred in the same way both inside a ship in uniform rectilinear motion and on the land. Therefore,

---

[4]E. N. VII, pp 212(32)–214(13). (Drake, pp 216–218).

[5]«When bodies are enclosed in a given space, their motions in relation to one another are the same whether the space is at rest or whether it is moving uniformly straight forward without circular motion.». See: Isaac Newton **The Principia**, op. cit. p. 423. Corollary V has been left unchanged in the three editions, in the form we have quoted.

implicitly, he maintained that a mechanical experiment could not disclose the state of motion or rest of the reference system. Huygens' collision theory was included in the work **De Motu corporum ex percussion** written in 1667, but published posthumous in 1703.[6]

However, some parts (among them the law of inertia) had been inserted in the **Horologium**.

---

[6]See: **Christiani Huygenii Opuscula Postuma**, Lugduni Patavorum—1703, pp 367-398.

Final Considerations

There is a risk of obviousness if we say that the adhesion to the Copernicanism and the consequent "Copernican militancy" have been for Galileo the causes which motivated and, we can say, directed his life for almost thirty years.

As we have already reminded, he began to consider positively the Copernican hypothesis in the Pisan period; he himself recalls this in the letter to Jacopo Mazzoni of 1597. Three important letters on the subject also belong to the same year.

Let us begin from that of Kepler to his former master Michael Maestlin (September 1597). Let us see the statement included:

«… Nuper in Italiam misi 2 exemplaria mei opusculi[1] (sive tui potius), quae gratissimo et lubentissimo animo accepit Paduanus Mathematicus, nomine Galilaeus Galilaeus, uti se subscripsit. Est enim et ipse in Copernicana haeresi inde a multis annis …»[2]

Therefore, Kepler communicates to Maestlin that Galileo has shared the "Copernican heresy" for many years. He does nothing other than to repeat what said by Galileo in the letter sent to him for thanking him for the free gift of a copy of the **Mysterium Cosmographicum**. In fact, Galileo writes (the fourth of August 1597):

«…. in Copernici sententiam multis abhinc annis venerim, ac ex tali positione multorum etiam naturalium effectuum caussae sint a me adinventae, quae dubio procul per comunem hypothesim inexplicabiles sunt. Multas conscripsi et rationes et argumentorum in contrarium eversiones, quas tamen in lucem hucusque proferre non sum ausus, fortuna

---

[1]It is the first Kepler's work: **Prodromus dissertationum cosmographicarum, continens mysterium cosmographicum de admirabili proportione orbium coelestium, …** (1596). For the sake of brevity, usually quoted as **Mysterium Cosmographicum**.

[2]E. N. X, p. 69. «… Recently, I sent in Italy two copies of my booklet (or better of yours), which were accepted with pleasure and gratitude by the Paduan mathematician, named Galileo Galilei, as he signed himself. Indeed, he has agreed to Copernican heresy for many years …» (Our translation).

© Springer International Publishing Switzerland 2016
D. Boccaletti, *Galileo and the Equations of Motion*,
DOI 10.1007/978-3-319-20134-4

ipsius Copernici, praeceptoris nostri, perterritus, qui, licet sibi apud aliquos immortalem famam paraverit, apud infinitos tamen (tantus enim est stultorum numerus) ridendus et explodendus prodiit.»[3]

It clearly comes out that, at that time, Galileo was not yet prepared to support the Copernican theory publicly, even if he proved to be convinced of it. In fact, Kepler's incitements (in the subsequent letter[4] of 13th October 1597) to perform accurate astronomical observations for confirming the Copernican theory had no effect. In the same letter, Kepler invites him anyhow to communicate him the results privately:

«Tu saltem scriptis mihi communica privatim, si publice non placet, si quid in Copernici commodum invenisti.»[5]

However, not even the observations regarding the *nova* of 1604 (until then Galileo had not performed astronomical observations) brought elements in favour of the Copernican theory and all seemed to lie over until the discovery and the use of the telescope.[6]

The first assertion of adhesion to Copernic's theory, as we would say nowadays "by press", however, appears only in the Postscript to the **Letters on the Sunspots (Istoria e Dimostrazioni intorno alle macchie solari)** of 1613.

The adhesion to the Copernican theory is expressed in indirect way, almost by coding:

«… Sono tali eclissi, ora di lunga durazione, ora di breve, e tal'ora invisibili a noi, e queste diversità nascono dal movimento annuo della terra, dalle diverse latitudini di Giove …»[7].

In December of that year it will follow the letter to Castelli and about two years later the famous letter to the Grand Duchess Cristina. Thus, there appear the first statements in support of the Copernican theory in a, so to say, semi-public form, that is in the shape of letters to particularly important persons.

---

[3]E. N. X, p. 68. «… I have agreed to Copernican theory for many years, and from such a position I have also discovered the causes of many natural effects which undoubtedly result inexplicable by following the common hypothesis. I have written many reasons and confutations of the contrary arguments that however until now I have not wanted to expose them publicly, frightened by the fortune of Copernicus himself, our master who, although has acquired an immortal reputation among a few, however has been disapproved and ridiculed by infinite others (so large is the number of blockheads).» (Our translation).

[4]E. N. X, pp 69–71.

[5]Ibidem. «If you should find something in favour of Copernicus, let me at least know it in writing, if you do not like to make it public». (Our translation).

[6]Regarding the period 1605–1609, there are not (at least) left letters, or something else, concerning the Copernican theory.

[7]E. N. V, p. 248. (Such eclipses, now of long duration, sometimes of short duration, and sometimes invisible to us, and these diversities originate from the annual motion of the earth, from the different latitudes of Jupiter…). (Our translation).

Generally, copies of those letters circulated with the consent of the author, except some cases in which the copies were passed through the hands of his opponents and had been modified on purpose.

This long preamble is for reaffirming, although there is no need, that Galileo's life after his comeback to Florence has been dominated by his "Copernican militancy" and the mechanics not always could find place in his thought with the necessary availability. One could object that, quite after his comeback to Florence, the year before the publication of the **Letters on the Sunspots**, Galileo had published what is considered his first book of physics, i.e. the already quoted **Discourse on the Bodies that stay atop water, or move in it (Discorso intorno alle cose, che stanno in sù l'acqua, ò che in quella si muovono)**.[8]

But, in that case, it was a dispute on subjects of hydrostatics in which Galileo had been reluctantly involved (against "philosophers", among whom particularly Ludovico Delle Colombe) and he had been explicitly requested to write it by the Grand Duke. He had not, instead, undertaken the work of "mechanics" he had announced in the letters to Antonio de' Medici and Belisario Vinta, of which we have already told.

Therefore, one can agree with Drake, when he says:

«… the dormant Copernican battle began to take shape at the end of 1613.»[9]

***

In the exposition of the subjects we have planned to deal with, we have given until now, we have striven for keeping, the most strictly possible, to the texts published by Galileo himself and to those left handwritten and published posthumous but of secure attribution.

Why do we insist on this? Because the study of the not-published, i.e. of the manuscript fragments (both those transferred by Favaro in the Edizione Nazionale, and those left in the folders of the National Library of Florence) conserved by Viviani and others, gave rise to various theories regarding the chronological sequence of them, contributing to support divergent interpretations of the results obtained by Galileo.

Anyway, one can be sure enough of some circumstances. Surely Galileo devoted his study to the mechanics in the Paduan period from 1602 to 1609; in 1602 he already planned a treatise on the natural motion. In this period he has also performed most of the experiments, getting constructed a laboratory for himself and having recruited a technical assistant. But after 1609, i.e. after the construction of the telescope, the studies of mechanics ceased and begun again only in 1618.

We have already noticed above the events occurred after his comeback to Florence. Later on, there will be the dispute with Lotario Sarsi (Orazio Grassi) and the publication of the Essayer (1623). The studies of mechanics will begin again in

---

[8]Published in two successive editions in the same year 1612 by Cosimo Giunti in Florence.
[9]Stillman Drake: **Cause Experiment & Science**—The University of Chicago Press, 1981—p. XVI. The volume also contains the English translation of the **Discorso...** of 1612.

two periods: 1626–27 and 1630–31. The results obtained in the last period will be also used in the Dialogue.[10]

At last, after the condemnation, unhealthy and, in the last years, also blind, Galileo devotes himself to the writing of the Discourses.

Thus, the treatise, dreamt since 1602, materialized after more than thirty years. We have already said that the Elzevirian edition of 1638 included only four Days of the six estimated. In addition, also the four published suffered from the difficulties in which Galileo was obliged to work for reasons of health and isolation.

Nevertheless, even in its bilingual "mixed" form, the work results to be clear in its basic statements. The study of the handwritten fragments, remnants of the preparatory phases, can certainly be useful to the specialists for better making clear the genesis of certain concepts and the time scansion of the variations for the final version, but cannot change it.

Anyhow, we remind the painstaking research, gone even to examine the quality of the paper and of the watermarks, made by Stillman Drake on the handwritten notes both of the Paduan period and the Florentine one.[11]

Obviously, as it is natural, also the theses of Drake have been brought into question, but this is not a matter in which we can go into.

Our purpose, as we have repeated many times, was that of checking if we can rightfully consider Galileo as the forefather of the first two laws of the Newtonian dynamics and therefore tout court of the dynamics, according to the statement of Ernst Mach.

Of course, it was a restricted purpose since we had not planned to investigate *in toto* the "mechanical" work of Galileo. We think of having fulfilled the duty and confirmed a positive conclusion to the *vexata quaestio* regarding the inertia principle.

Certainly the statements, earlier by Duhem and successively by Koyré, have heavyly laid on the scholars of Galileo for a long time. The influence of the latter who wanted Galileo as a "Platonic philosopher" still survives.

But the revival of Galileo founder of the experimental method (because just with a revival we have to do, since his talents of experimenter were largely recognized by the contemporaries) is anyway actually on.

---

[10]We do not agree with Clavelin who holds the existence of: «... différénces, pourtant évidentes, qui séparent le Dialogue et les Discours.» (M. Clavelin, op.cit. p. 182). It is true, as Clavelin points out, that in the Discourses the principle of relativity is no more reminded, but the law of fall of the heavy bodies, for instance, is reported in the Dialogue in the same form as in the Discourses.

[11]See: Stillman Drake—**On the probable order of Galileo's notes on motion**—Physis, 14, 55-68, 1972; **Galileo's experimental confirmation of horizontal inertia: unpublished manuscripts**—Isis, 64, 290- 305, 1973; **Galileo's notes on motion—arranged in probable order of composition and presented in reduced facsimile**—Supplemento agli Annali dell'Istituto e Museo di Storia della Scienza, Anno 1979. Fascicolo 2.

Name Index

Note: Galileo's name is not included

A
Albèri, E., 63, 64
Albert of Saxony, 36, 39, 47, 56
Albertus Magnus, 31, 40
Alighieri (Dante), xiii, 56
Ambrogetti, M., 111
Anaximander, 3
Anaximenes, 3
Angel of Fossanbruno, 43
Apollonius, 153
Aquinas (St.). T., 31, 87
Archimedes, 4, 9, 10, 14, 20, 21, 31, 44, 50,
        53, 64, 68, 70, 71, 88, 104, 106
Aristarchus of Samos, 164
Aristotle, 4, 5, 8, 11, 14–16, 18, 20–22, 25,
        26, 30, 32, 38, 44, 48, 52, 53, 55, 57,
        67–72, 75–78, 85, 87, 93, 118, 148, 164
Arnold, V., xi
Atwood, G., 113
Autolicus of Pitane, 9
Avempace, 30, 69, 85, 87, 93
Averroes, 30, 38, 87
Avicenna, 31, 87

B
Baldi, B., 26, 44
Baliani,G.B., 141, 160
Bell, E.T., 10
Benedetti, G.B., 25, 44, 48, 50, 52, 96
Biagio (Pelacani), 42, 43
Blackwell, R.J., 116
Bonamici, F., 88

Bordiga, G., 52
Borel, G., 7, 11
Borsetto, L., 72
Bradwardine, T., 26, 32, 34, 38, 39
Brunelleschi, F., 58
Burckhardt, J., 44
Buridan, J., xii, 26, 36, 39–41, 43, 56, 88, 164

C
Camerota, M., 64
Capra, G.P., 56
Cardano, G., 44, 45, 79, 96
Castelli, B., 101, 142
Cavalieri, B., 124, 125, 151
Caverni, R., 28, 50, 55–57, 88, 127
Chrisippus of Soli, 19
Chwolson, O.D., 113
Clagett, M., 31–33, 35
Clavelin, M., 86, 102
Clavius, C., 61
Cleanthes of Asso, 19
Cleomedes, 20
Coffa, J.A., 102, 104, 105, 111
Cohen, B., 127, 128
Commandino, F., 44, 79
Copernicus, N., 57, 75
Cozzi, L., 150
Cusanus (Nicholas of Cusa), 96, 165

D
Damerow, P., 149, 161
Danesi, L., 90
De'Medici, A., 101, 150, 171
De Pace, A., 57

© Springer International Publishing Switzerland 2016
D. Boccaletti, *Galileo and the Equations of Motion*,
DOI 10.1007/978-3-319-20134-4

De Santillana, G., 16
Del Lungo, I., 64
Del Monte, G., 44, 57, 90, 96, 125, 148, 149,
    155
Delle Colombe, L., 171
Democritus of Abdera, 3
Denomy, A.J., 165
Descartes, R., xiv, 106, 115, 128, 139, 153,
    166
di Giorgio Martini, F., 58
di Marchia, F., 40
Dostoevskij, F., xiii
Drabkin, I.E., xiv, 64, 74
Drake, S., xiv, 93, 94, 96, 108, 127, 128, 140
Dugas, R., 27
Duhem, P., xii, 26–28, 39, 55, 86
Dumbleton, J., 33, 35
Duns Scoto, 87
Dürer, A., 58

E
Elzevir, L., 111
Eneström, G., 31
Enriques, F., 28
Epicurus, 18
Eratostenes, 20
Euclid, 9, 31, 44–46, 51, 53, 153
Eudoxus, 8
Euler, L., 10

F
Favaro, A., 28, 64
Federici Vescovini, G., 42
Ferrini, M.F., 82
Finocchiaro, M.A., 128
Frank, M., 149
Fredette, R., 64, 74, 79, 88
Frisi, P., 155

G
Galluzzi, P., 83, 89, 90
Gassendi, P., 106, 115, 160
Gatto, R., 79, 80, 89
Gaurico, L., 51
Gaye, R.K., 4
Gerard of Brussels, 31, 32
Gerard of Cremona, 25
Ghiberti, L., 58
Gilbert, W., 115
Gille, B., 58
Gillispie, Ch.C., 36

Giusti, E., 31, 53, 54, 65, 137
Goulet, R., 20
Grant, E., 34, 36
Grassi, O.(Sarsi, L.), 101, 171
Grosseteste, R., 25
Guzman, G., 52

H
Hardie, R.P., 4
Heath, T.L., 10
Hero of Alexandria, 18, 20, 21, 44, 79
Heytesbury, W., 33–35, 43
Hooper, W.E., 64
Hopkins, J., 96, 165
Huygens, C., 116, 155, 166

J
Jacopus of Forlì, 43
Jordanus of Nemore, 44, 81
Jowet, B., 5

K
Kepler, J., 115
Koestler, A., 118
Koyré, A., 11, 55, 64, 94, 102, 106, 108, 109,
    115, 127, 128, 140, 161

L
Lamar Crosby, H., 33
Leonard of Vinci, xii, 44
Lewis, J.T., 43
Libri, G., 50, 88, 149
Loria, G., 9
Loyd, G.E.R., 16, 17
Lucretius Carus, T., 19

M
Maccagni, C., 53
Mach, E., xii, 27, 94, 127
Maestlin, M., 169
Magalotti, L., 66
Maier, A., 27, 39
Marcolongo, R., 28
Marsili, C., 125, 149
Matteoli, S., 57
Mazur, J., 32
Mazzoni, J., 56, 57, 76, 88
Mc Cormac, T.J., 27
Menut, A.D., 165

Mersenne, M., xiv, 90, 91, 153, 160, 161
Messisbugo, C., 67
Mieli, A., 28
Moerbeke (of), W., 25
Moody, E.A., 67, 86

N
Naylor, R.H., 128, 161
Nelli (Clementi), G.B., 67
Nenci, E., 26
Newton, I., vii, 10, 94, 166

O
Oddi, M., 125
Olschki, L., 64
Oresme, N., xii, 35–37, 39, 43, 47, 164

P
Palmieri, P., 161
Panzanini, J., 66
Pappus of Alexandria, 99
Parmenides, 5
Pelacani (Biagio), 42, 43
Philoponus, J., 29, 38, 39, 69, 87, 93
Plato, 5, 57
Pliny the Elder, 20
Posidonius of Apameia, 19
Procissi, A., 67
Ptolemy, C., 20, 68, 75
Pythagoras, 3

R
Renn, J., 149, 161
Rieger, S., 149, 161
Roberval (de), G. P., 160
Rossi, P., xiii
Russel, B., 5, 32

S
Saint-Vincent (de), G.., 5
Sambursky, S., 21

Sargent, S.D., 27
Sarpi, P., 101, 128, 137, 150
Scheiner, C., 99, 100, 122
Scot, M., 25
Seneca, L.A., 20
Settle, T.B., 64, 128, 139
Shakespeare, W., xiii
Simplicius, 17
Sosio, L., 150
Souffrin, P., 8
Speroni, S., 59
Stevin, S., 52
Stocks, J. L., 4
Strato of Lampsacus, 17

T
Taisner, G., 52
Targioni Tozzetti, G., 66, 67
Tartaglia, N., 25, 44, 45, 47, 50, 51, 149
Thales of Miletus, 3
Theophrastus, 17
Torricelli, E., 110, 147, 154, 160

V
Vailati, G., 27, 88
Ventrice, P., 52
Venturi, G.B., 67
Vergara Caffarelli, R., 111, 113
Vinta, B., 101, 151
Viviani, V., 65, 142

W
Welser, M., 99
Wohlwill, E., 27, 55, 64, 127
Wilson, C.A., 33
Wisan, W., 127

Z
Zeno of Citium, 19
Zeno of Elea, 5

Printed in the United States
By Bookmasters